EVOLUTION AND STRUCTURE OF THE INTERNET
A Statistical Physics Approach

This book describes the application of statistical physics and complex systems theory to the study of the evolution and structure of the Internet.

Using a statistical physics approach, the Internet is viewed as a growing system that evolves in time through the addition and removal of nodes and links. This perspective permits us to outline the dynamical theory required for a description of the macroscopic evolution of the Internet. The presence of such a theoretical framework appears to be a revolutionary and promising path towards our understanding of the Internet and the various processes taking place on this network, including, for example, the spread of computer viruses or resilience to random or intentional damages.

The presentation focuses on statistical regularities observed in the large-scale structure of the network, the so-called "global Internet" as well as on the importance of dynamics in the formulation of adequate models. Using this approach it is possible to provide a unified picture of results obtained on the Internet in the context of different scientific communities. This makes use of methods and concepts that have proven to be extremely useful in the analysis of more classical statistical physics systems, such as percolation theory, mean-field methods, and cellular automata simulations.

This book will be of interest to graduate students and researchers in statistical physics, computer science, and mathematics studying the structure and evolution of the internet.

ROMUALDO PASTOR-SATORRAS received his Ph.D. at the University of Barcelona. He has been a research fellow at Yale University and at the Massachusetts Institute of Technology in Cambridge, MA. He spent two years as a research fellow at the International Center for Theoretical Physics (UNESCO) and then moved back to Spain in 2000 as Assistant Professor at the University of Barcelona. Since 2001, Pastor-Satorras has been a research scientist and lecturer at the Universitat Politècnica de Catalunya. He is the author of more than 40 research papers in different areas of non-equilibrium statistical physics, condensed matter theory, and complex systems analysis.

ALESSANDRO VESPIGNANI obtained his Ph.D. at the University of Rome "La Sapienza." After holding research positions at the Universities of Yale and Leiden, he joined the condensed matter research group at the International Center for Theoretical Physics (UNESCO) in Trieste where he has carried out research and teaching activities for more than five years. He has authored more than 100 scientific papers on the statistical physics of non-equilibrium phenomena, critical phase transitions, and complex and disordered systems. At present he is Senior Research Scientist of the French National Council of Scientific Research (CNRS), at the Université de Paris-Sud, France.

EVOLUTION AND STRUCTURE
OF THE INTERNET
A Statistical Physics Approach

ROMUALDO PASTOR-SATORRAS

Universitat Politécnica de Catalunya
Barcelona, Spain

ALESSANDRO VESPIGNANI

Laboratoire de Physique Théorique, Université de Paris-Sud
Orsay, France

CAMBRIDGE
UNIVERSITY PRESS

CAMBRIDGE UNIVERSITY PRESS
Cambridge, New York, Melbourne, Madrid, Cape Town, Singapore, São Paulo

Cambridge University Press
The Edinburgh Building, Cambridge CB2 8RU, UK

Published in the United States of America by Cambridge University Press, New York

www.cambridge.org
Information on this title: www.cambridge.org/9780521826983

First published 2004
This digitally printed version 2007

A catalogue record for this publication is available from the British Library

Library of Congress Cataloguing in Publication data

Pastor-Satorras, R. (Romualdo), 1967–
Evolution and structure of the Internet: a statistical physics approach / Romualdo Pastor-Satorras,
Alessandro Vespignani.
p.   cm.
Includes bibliographical references and index.
ISBN 0 521 82698 5
1. Internet.   2. Statistical physics.   I. Vespignani, Alessandro, 1965–   II. Title.
TK5105.875.I57P385 2004
004.67′8–dc21   2003047257

ISBN 978-0-521-82698-3 hardback
ISBN 978-0-521-71477-8 paperback

Contents

Preface

For the majority of people the word "Internet" means access to an e-mail account and the ability to mine data through any one of the most popular public web search engines. The Internet, however, is much more than that. In simple terms, it is a physical system that can be defined as a collection of independently administered computer networks, each one of them (providers, academic and governmental institutions, private companies, etc.) having its own administration, rules, and policies. There is no central authority overseeing the growth of this networks-of-networks, where new connection lines (links) and computers (nodes) are being added on a daily basis. Therefore, while conceived by human design, the Internet can be considered as a prominent example of a self-organized system that combines human associative capabilities and technical skills.

The exponential growth of this network has led many researchers to realize that a scientific understanding of the Internet is necessarily related to the mathematical and physical characterization of its structure. Drawing a map of the Internet's physical architecture is the natural starting point for this enterprise, and various research projects have been devoted to collecting data on Internet nodes and their physical connections. The result of this effort has been the construction of graph-like representations of large portions of the Internet. The statistical analysis of these maps has highlighted, to the surprise of many, a very complex and heterogeneous topology with statistical fluctuations extending over many scale lengths. These are the typical signatures of emergent phenomena, as we observe in nature in many complex systems which are subject to dynamical evolution.

When looking at networks from the point of view of complex systems, the focus is placed on the microscopic processes that rule the appearance and disappearance of nodes and links. Then, since the system is composed of very many interacting units, a detailed evaluation of the dynamics of each unit is avoided in favor of the understanding of the cooperative phenomena originated by their dynamical interactions and the statistical laws governing the system. Such a methodology is akin to the statistical physics approach that has been successfully applied to link the microscopic dynamics and interactions of atoms and matter to the statistical regularities of macroscopic physical systems.

Following this path, an intense research activity has been devoted in recent times to apply the statistical physics approach to the study of complex growing networks in general, and the Internet in particular. In the statistical physics approach the Internet is viewed as a growing system that evolves in time by adding and removing nodes and links. This perspective is somehow opposite to the traditional static graph modeling and allows the identification of some basic models that, while still missing many details, appear to outline the dynamical theory required for the description of the macroscopic Internet's evolution. The presence of such a theoretical framework appears as a revolutionary and promising path in our understanding of the Internet and other complex technological and natural networks.

The introduction of the statistical physics approach into the field of network studies has also provided new techniques and methods with which to approach problems related to network topology, such as resilience to damage and diffusion or searching processes. In this case, well-established techniques in statistical physics, such as percolation theory, mean-field methods, cellular automata simulations, etc., can be used to gain a deeper understanding of the Internet's properties.

The purpose of this book is to provide a unified picture of the results obtained about the Internet in the context of different scientific communities by privileging the use of methods and concepts that have proven to be extremely useful in the analysis of more classical statistical physics systems. We shall therefore make a strong emphasis on the statistical regularities observed in the large-scale structure of the network, the so-called global Internet, and the importance of the dynamics in the formulation of adequate models. In doing this, we have made a special effort to bridge the language gap that might occur among different communities by devoting the two initial chapters to an outline of the Internet's history and an elementary description of its functioning. This will allow us to build up a basic Internet glossary and outline the main elements that make the Internet work. We also provide an appendix summarizing the main concepts of graph theory, which are used in the topological description of Internet maps.

The road map of the book can be schematized in two main parts. The first six chapters are essentially devoted to the physical Internet. In these chapters we review the various experimental projects dealing with data collection, focusing on the various mapping strategies and the level of description achieved with different tools. Following this, we present the statistical analysis of the most recent data available, discussing in detail the main topological features characterizing the Internet's large-scale topology. The ensuing chapter contains an overview of models which propose to represent the Internet. Here we emphasize the "physicist" point of view by introducing the reader to the modern field of growing network models. Finally, we report in Chapter 6 an analysis of the Internet's resilience to damage by casting the problem in the general framework of phase transitions and percolation phenomena.

The second part consisting of Chapters 7, 8, and 9 is instead focused on the virtual networks hosted by the Internet, such as the World Wide Web, peer-to-peer systems, and other social communities, and to dynamical phenomena that occur on them, such as search processes and epidemic spreading. Finally, Chapter 10 is a short discussion of important features that are likely going to represent the main challenges for a full understanding of the Internet in the near future.

The systematic study of the large-scale properties of the Internet and its view as a complex evolving network, while a relatively recent field, has generated quite a large number of works and a vast literature on the subject. We have made every effort to account and mention all the works relevant for a proper understanding of each chapter. It is, however, quite impossible to discuss in detail all the contributions to the field and we have therefore made some choices based on our perception of what is more relevant to the focus of the present book. We apologize to all the colleagues who feel that their specific contributions have been overlooked. We hope that our effort will result in a comprehensive and useful presentation of the subject to everybody working in the field, and, more specially, to any researcher or student who intends to enter it. In this sense, by conveying the idea that the Internet is a paradigmatic example of complex system, we believe that the book can be of interest to computer scientists, physicists, and mathematicians alike.

Many people have contributed to the preparation of this book, specially by shaping our own understanding of the subject. Most of what we know about the Internet is the result of past and present scientific collaborations with M. Boguñá, G. Caldarelli, Y. Moreno, R. Percacci, R. V. Solé, and A. Vázquez. Many of the subtle technicalities of the Internet working has been explained to us by I. Alvarez-Hamelin and S. Visintin to whom goes our deepest gratitude. Our views and knowledge of complex networks have been refined through invaluable scientific discussions with L. Adamic, I. Alvarez-Hamelin, A.-L. Barabási, A. Barrat, G. Bianconi, C. Castellano, A. Flammini, A. Krzywicki, S. Havlin, M. Latapy, A. Lloyd, R. May, F. Menczer, A. Maritan, M. E. J. Newman, R. Percacci, L. Pietronero, N. Schabanel, R. V. Solé and F. van Wijland. The first drafts of the book were read and criticized by I. Alvarez-Hamelin, A.-L. Barabási, A. Barrat, M. Barthelemy, G. Caldarelli, L. Fabbian, A. Flammini, D. Iaschi, C. Magnien, F. Menczer, M.-C. Miguel, Y. Moreno, T. Pernice, A. Vázquez, and M. Vergassola. The book would look very different, and much worse, without their thorough revisions and comments. Many of the figures and plots reported in the book would not be there without the kind help and raw data retrieval of G. Caldarelli, G. Bianconi, S. Bornholdt, H. Ebel, H. Jeong, S. Leonardi, E. Ravasz and A. Vázquez. Particularly, we want to express our warmest gratitude to A.-L. Barabási for the encouragement and the warm hospitality in the University of Notre Dame, where parts of the book were written. We are also grateful to the friends at the International Center for Theoretical Physics in Trieste, where the book writing was initiated. We wish to thank the constant editorial guidance of Simon Capelin and the continuous help and support of the outstanding editorial staff at Cambridge University Press. We also thank the various institutions which have made possible our work through their generous support at the various stages of this project: the FET Open project "COSIN" IST-2001-33555, the Ministerio de Ciencia y Tecnología (Spain) through its program "Ramón y Cajal" and grant BFM 2001–2154, the Action Specifique CNRS/STIC *Dynamo* and the BQR *Graphes du Web* of Orsay University.

Romualdo Pastor-Satorras
Alessandro Vespignani

Abbreviations

APNIC	Asia Pacific Network Information Centre
ARIN	American Registry for Internet Numbers
ARPA	Advanced Research Project Agency
ARPANET	ARPA Network
AS	Autonomous System
BBN	Bolt, Beranek, and Newman Corporation
BGP	Border Gateway Protocol
BITNET	Because It's Time Network
CAIDA	Cooperative Association for Internet Data Analysis
CERN	Counseil Europeen pour la Recherche Nucleaire
CPU	Central Processing Unit
CSNET	Computer Science Research Network
DARPA	Defense Advanced Research Project Agency
DC	Disconnected Components
DNS	Domain Name System
EGP	Exterior Gateway Protocol
FTP	File Transmission Protocol
GIN	Giant In-Component
GOUT	Giant Out-Component
GSCC	Giant Strongly Connected Component
GWCC	Giant Weakly Connected Component
HEPNET	High Energy Physics Research Network
HOT	Heuristically Optimized Trade-off
HTML	HyperText Markup Language
HTTP	HyperText Transfer Protocol
ICMP	Internet Control Message Protocol

IESG	Internet Steering Group
IETF	Internet Engineering Task Force
IGP	Interior Gateway Protocol
IMP	Interface Message Processors
INWG	Internetworking Working Group
IP	Internet Protocol
IPMA	Internet Performance Measurement and Analysis Project
IR	Internet Router
IRR	Internet Routing Registry
ISO	International Organization for Standardization
ISP	Internet Service Provider
LACNIC	Latin American and Caribbean IP address Regional Registry
LAN	Local Area Network
MAN	Metropolitan Area Network
MFENET	Magnetic Fusion Energy Research Nerwork
MILNET	Military Network
MIT	Massachusetts Institute of Technology
NASA	National Aeronautics and Space Administration
NCP	Network Control Protocol
NLANR	National Laboratory for Applied Network Research
NSF	National Science Foundation
NSFNET	NSF Network
OSI	Open System Interconnection
P2P	Peer-to-Peer system
RAM	Random Access Memory
RFC	Request for Comments
RIPE	Réseaux IP Européens
RIP	Routing Information Protocol
RTT	round-trip-time
SIR	Susceptible-Infected-Removed model
SIS	Susceptible-Infected-Susceptible model
SMTP	Simple Mail Transfer Protocol
SPAN	Space Physics Analysis Network
SRI	Stanford Research Institute
TCP	Transmission Control Protocol
TTL	time-to-live
UCLA	University of California at Los Angeles

UDP	User Datagram Protocol
URL	Uniform Resource Locator
USENET	Unix User Network
UUCP	Unix User Control Protocol
WAN	Wide Area Network
WWW	World Wide Web

1

A brief history of the Internet

The Internet is the result of the bold effort of a group of people in the 1960s, who foresaw the great potential of a computer-based communication system to share scientific and research information. While in the early times it was not a user-friendly environment and was only used by a restricted community of computer experts and scientists, nowadays the Internet connects more than one hundred million hosts[1] and keeps growing at a pace unknown in any other communication media. From this perspective, the Internet can be considered as one of the most representative accomplishments of sustained investment in research at both the basic and applied science levels.

The success of the Internet is due to its world-wide broadcasting capability that allows the interaction between individuals without regard for geographic location and distance. The information exchanged between computers is divided into data packets and sent to special devices, called *routers*, that transfer the packets across the Internet's different networks. Of course a router is not linked to every other router. It just decides on the direction the data packets take. In order to work reliably on a global scale, such a network of networks must be very slightly affected by local or even extensive failures in the network's nodes. This means that if a site is not working properly or it is too slow, data packets can be rerouted, on the spot, somewhere else. Surprisingly, such a network communication system is realized by a complex interplay of communication protocols, hardware infrastructures, and connectivity architecture that is the outcome of an evolution lacking any central authority. Things have always been changing on the Internet, sometimes gradually and sometimes very rapidly, but always evolving without a precise general design. The Internet is in this sense a major example of a *self-organizing system*, combining human needs and technological capabilities in a cooperative way.

---

[1] A host is defined as a computer that allows individual users to communicate with other computers through the Internet.

In this chapter we want to provide a brief history of the Internet's development and growth, introducing at the same time the basic terminology and concepts used to describe this system. The main scope of the following pages is to allow the reader to understand the complex nature of the evolution of the Internet, and to highlight its most important technological and organizational ingredients. From this perspective, and for the sake of simplicity, the present chapter is necessarily a sketchy presentation of the Internet's history. The interested reader can dig deeper into this subject in the works of Gillies and Cailliau (2000), Abbate (2000), the Hobbes' Internet Timeline (2000)[2], and the online history by Leiner *et al.* (2000)[3].

1.1 The early times

1.1.1 Distributed communication networks

The history of the Internet began in the early 1960s, at the height of the cold war, when ARPA,[4] an agency created in 1958 by the US Department of Defense to sponsor research projects related to military problems, started to fund programs at universities and corporations concerning the creation of a computer network to access and share data and programs among computers located in different places. This was an appealing project, specially from the perspective of providing secure control over information in the event of large-scale international conflicts.[5]

It is important to recall that at that time computer communications were only point to point, with each network link depending upon the link before it. In such a structure, any part of the system could easily be isolated by knocking out just one of the links. A different and obvious topology for a computer network in the 1960's was a highly centralized one, in which all computers were connected to a central unit handling all data exchanges, the so-called "star-shape" topology. Those centralized systems, however, are even more vulnerable to attacks and failures, since knocking out the central node is enough to disconnect the whole network.

It is possible to devise several other regular topologies, such as those represented in Figure 1.1, though their respective vulnerability to external attacks or internal failures was uknown. With this prospect at hand, Paul Baran at RAND corporation was given a US Air Force grant to investigate how the US military could protect its communication systems from serious damage. In his conclusions, Baran outlined the principle of "redundancy of connectivity" and explored various models for designing communication systems and evaluating their vulnerability. In the ensuing

---

[2] http://www.zakon.org/robert/internet/timeline/
[3] http://www.isoc.org/internet/history/brief.shtml.
[4] Advanced Research Project Agency. ARPA was renamed DARPA, the Defense Advanced Research Project Agency, in 1972.
[5] Despite the fact that the later studies on survivability and robustness considered explicitly the nuclear threat (Baran, 1964), the initial ARPA was not officially related to building a network resistant to nuclear war.

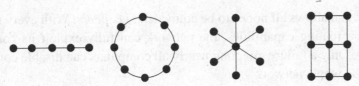

Fig. 1.1 Examples of regular topologies: from left to right, linear, ring, star, and mesh.

report, Baran (1964) proposed a distributed communication system, in which there would be no obvious central unit, every node having the same routing capabilities. With such a configuration, all points surviving a large-scale attack would be able to maintain the contact with the surviving part of the network.

The best design for such a decentralized network is obviously a highly inter-connected, distributed network in which each node is connected to all the others. This is a fully connected network with the highest degree of redundancy. The first plan was indeed to connect the mainframe computers at each site directly to all others, but it was soon realized that such a level of redundancy was far too compli-cated and expensive to be handled. A different approach was thus pursued, design-ing a *distributed network* with sufficient redundancy, but far from a fully connected topology.

1.1.2 Packet switching technology

The technology capable to handle communications in a distributed network was provided by the work on data packet switches initiated by several groups in the mid-1960s.[6] The group led by Frank Heart, at the Bolt, Beranek, and Newman (BBN) corporation, was awarded, by ARPA in 1968, a contract to develop the Interface Message Processors (IMPs), small machines – the ancestors of modern routers – which were designed to be a part of each mainframe dedicated to form the subnetwork between computers. The IMPs used a technology called *packet-switching*, which parcels data in small chunks called packets, labeled with the des-tination address. Packets can be sent in any order through any path leading to the same destination and reassembled on arrival. The advantage of a packet-switching system is evident: in a centralized network, such as a star-shaped system, all the information is channeled to the central unit, to be processed and redistributed. In a distributed network with packet-switching technology, however, each node has the authority to originate, pass, and receive messages. In particular, if a node is not working, or working too slowly, a packet can be rerouted through some other nodes. This dynamical rerouting implies that each packet finds its own way through

---

[6] The first report on packet switching theory, dated in 1961, was written by Leonard Kleinrock while a graduate student at MIT.

the network and allows all nodes to be equivalent, i.e. *peers*. With every node having the same routing capabilities, the network can fully exploit its connectivity redundancy: Only a failure affecting nearly all computers can disable communications over the whole network.

1.1.3 The ARPANET

In 1966 the MIT researcher Lawrence G. Roberts started for ARPA the design of ARPANET, a network initially intended to wire up four major mainframes at universities in Southwestern US: the University of California at Los Angeles (UCLA), the University of California at Santa Barbara, the Stanford Research Institute (SRI), and the University of Utah. In 1968 the ARPANET specifications were laid down, based on the IMP technology, and the first host-to-host message was sent from UCLA to SRI in October 1969. The four planned mainframes were connected by the end of 1969, forming ARPANET, the precursor of the present Internet.

During the 1970–71 period, ARPANET grew to 23 nodes, while work proceeded quickly on designing a functionally host-to-host protocol called Network Control Protocol (NCP), which became fully implemented in 1971–72. Once a reliable working protocol was established, researchers started to develop applications, and in 1972 Ray Tomlison at BBN wrote the basic e-mail message software, which with time has become probably the most widely used application on the network. The same year the Internetworking Working Group (INWG) became the first standard-setting entity to govern the growing network. In 1973 the first international nodes were set up in England and Norway. In 1977 the ARPANET encompassed 107 nodes. Its growth was slow but steady, the scientific community recognizing that the new communication network was going to become something wider than they had ever imagined. This fact called for designs and technologies intended for a larger network and the Internet took off.

1.2 The rapid growth

1.2.1 More networks

In a short time, the ARPANET example was absorbed by other communities, such as the US Department of Energy, which established the MFENET for its researchers in magnetic fusion energy, and the HEPNET for high energy physicists. NASA space physicists followed with SPAN, enlarging the number of purpose-built networks for academic and research communities. Private companies soon also made their first move into the electronic world when BBN opened in 1975 Telenet, a commercial network based on the ARPANET model.

At the same time, other networking technologies were being developed by the computer science community. One such alternative was the CSNET (Computer Science Research Network), providing networks for computer science departments. CSNET benefited specially from the dissemination of the Unix operative system and its built-in Unix User Control Protocol (UUCP). In the early 1980s other important networks based on this technology, such as USENET (Unix User Network)[7] and BITNET (Because It's Time Network), sprang up. The reason for the surge in UUCP-based networks was that UUCP, modems, and the existing telephone lines were a ready-to-use technology of data transport. Also, computer facilities and institutions which were not part of the ARPANET were increasingly aware of the benefits of belonging to a large linked computer system. UUCP networks were, however, rather different than the ARPANET, providing essentially store-and-forward facilities and e-mail, in order to create discussion groups and provide two-way and one-to-many communication. In other words, the user's computer connects to another machine which has filed users' postings in separate topical discussion groups (newsgroups). The user may issue a command requesting the full text of a particular posting, post a follow up or even start a new newsgroup. These e-mail discussion lists constituted another major element in the building of the Internet community, heavily contributing to the international growth of the internetworking principles.

1.2.2 The TCP/IP development

The launch and growth of several different networks was a further stimulus for the development of the key technical idea underlying the modern Internet, namely that of *open architecture* networking. In an open architecture network, the individual networks may be designed in accordance with specific requirements that can be freely selected by each administration entity. Each network, however, communicates with the other networks through a set of protocols that are the same regardless of which network the user or service operates in. In other words, each network stands on its own and no internal changes are required to connect it to the Internet. In this respect, the NCP communication protocol used in ARPANET was not a viable solution, since it did not deal with end-to-end host errors and destinations different than the IMPs on the ARPANET. The first effort to develop an open-architecture network was led by Robert E. Kahn at BBN and Vinton G. Cerf at Stanford. Their research yielded as a final result the TCP/IP protocol. The TCP,

---

[7] USENET was developed by T. Truscott and J. Ellis while graduate students at Duke University and North Carolina University, respectively. At first it was just a graduate students activity but eventually grew until accommodating a link to join the ARPANET mailing list. Doubtless, USENET has been one of the main factors driving the physical and social self-organized nature of the Internet.

or *Transmission Control Protocol*, converts messages into a bunch of packets at the source and reassembles them back into messages at the destination. The IP, or *Internet Protocol*, however, handles the addressing and routing of single packets across nodes and different networks, providing a unique address space for the Internet. The TCP/IP was adopted as a standard of the US Department of Defense and was shared by other agencies, and in 1983 ARPANET experienced a complete transition from NCP to TCP/IP. The new protocol allowed the original ARPANET to split into two different networks, its military part, MILNET, and an ARPANET supporting only research needs. At the same time the TCP/IP, which was public-domain software, was adopted by other networks, such as the CSNET, to route information exchange with the ARPANET. As the use of the TCP/IP became more and more common, other entire networks began to interconnect and the network of networks was finally born.

1.2.3 The NSF acceleration

Another turning point in the history of the Internet was the National Science Foundation (NSF) program to establish a new transcontinental network and five super-computing centers. The program was driven by the idea that networking and computer resources were indispensable tools for the research community, and the NSF philosophy was to serve the entire higher education community, regardless of discipline. In order to allow all institutions to link up to the network, NSF agreed to pay for the establishment of a connection to its high-speed network (the *backbone*) only if universities provided connections to smaller educational and research institutions. A critical decision of the NSFNET (NSF Network) program was the mandatory use of the TCP/IP protocol, triggering the marginalization of other wide-area network protocols. Another NSF decision, which contributed to the shape of today's Internet, was to encourage commercial network traffic at the local/regional level along with the ban of the commercial use of the NSFNET backbone. This stimulated the establishment of long-haul private networks (national-scale providers) offering alternative backbones for the commercial traffic.

The NSFNET had an enormous impact on the Internet's evolution. The massive funding and the establishment of policy guidelines triggered the transition from a small network limited to the research environment to a full-scale network of networks with solids links also in the commercial community. The hardware technology received an impressive thrust as well, and the TCP/IP was put on its way to becoming the standard of the present. NSFNET co-opted the ARPANET in 1989 and finally, in 1995, reverted back to a pure research network, leaving the national-scale connectivity to private backbone providers.

1.3 The network of networks: a growing self-organized system

1.3.1 A large-scale infrastructure

From 1990 the Internet has experienced explosive growth. The number of hosts nowadays are counted in tens of millions and new networks and providers are connecting to the Internet on a daily basis (see Figure 1.2). The reasons for this phenomenon are partly due to the fact that personal computers have become an household item, and partly to the advent of the World Wide Web (see Chapter 7), which allows easy access to the huge amount of information stored in the Internet. This increase in scale of the Internet necessarily introduced several new concepts and changes in the underlying technologies. The shift from a few networks with a modest number of hosts to a large number of networks with a wide range of connected hosts has led to the definition of three possible classes of networks. Class A represents large national-scale networks (there are a few of these networks and they have a large number of hosts). Class B networks are regional-scale networks. Class C comprises small local area networks with a limited number of hosts.[8] Since hosts have a unique numeric address, a related name is assigned to each of them, to make it easier for people handling the addresses.

However, with millions of hosts it is impossible to have a single table of all the hosts and their associated names and addresses. This has led to the introduction of the Domain Name System (DNS) that allows the resolution of host names into

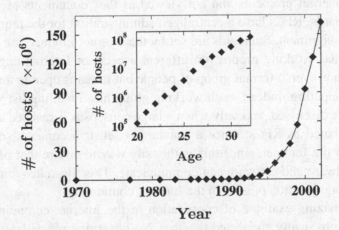

Fig. 1.2   Number of hosts in the Internet starting from 1970. The inset reports the number of hosts as a function of the Internet's age in a logarithmic-linear scale. The linear behavior shows that in the last ten years of life the Internet has grown exponentially, doubling annually. Data from the the Hobbes' Internet Timeline http://www.zakon.org/robert/internet/timeline/.

---

[8] The class subdivision results in IP addresses allowing different numbers of networks and related hosts, as explained in Chapter 2.

Internet addresses. The same sort of problem has occurred also for routers. Originally, all routers implemented a single distributed algorithm for the routing of data packets (a long list of paths to all addresses). As the number of networks grew, a hierarchical model of routing split the protocol into an Interior Gateway Protocol (IGP), used inside Internet regions, and an External Gateway Protocol (EGP) used to link different regions. With time, new problems related to Internet scalability are faced. New ideas continue to pop up as well, and new solutions to be envisioned. The Internet behaves in this sense as an evolving system, whose appearance is changing over the various phases of its life.

1.3.2 Self-organization and cooperation

During its growth the Internet has changed continuously, surviving dramatic changes in technology and evolving to accommodate the exponential increase in users. A key element for these "adaptive" features of the Internet is the dynamic exchange of ideas and the open access to technical standards. A clear example of this attitude is the Request For Comments (RFC) series of notes.[9] Originally these memos were intended to be a fast way to share and distribute ideas, with online files accessible via the File Transfer Protocol (FTP). The very early RFCs presented ideas developed by groups of researchers to the rest of the community, or information on protocols and engineering issues. Nowadays, RFCs are more focused on Internet protocols and are viewed as the "documents of record" for Internet standards. RFCs have a centralized administration for the required protocol number assignment. Standards are set by the Internet Engineering Task Force (IETF), that has working groups on different aspects of Internet engineering.[10] IETF is not, however, a formal group of people but remains open to anyone interested in participating. Indeed, each working group has a mailing list where draft documents are discussed, and only when a large consensus is reached are the documents distributed as RFCs. Once a standard is set, it becomes a sort of commandment on the Internet, since this is the only way to ensure that people using different hardware and software can communicate. This illustrates fairly well the level of self-organization present in the Internet community.

Another amazing example of cooperation in the Internet community is how networks are physically connected together. No governing office determines how routers must be connected. Each network organization, from small campuses to large national-scale providers, decides autonomously its connections and makes arrangements to pass along other's network traffic. From this perspective, it is natural

---

[9] http://www.rfc-editor.org/.

[10] The IETF has grown in time and resulted in further substructure, such as the Internet Steering Group (IESG) formed by working group directors.

that non-functional network connections (excessive loads, spam e-mail, denied services) are in a short time cut-off by their peers. In other words, a working Internet requires network managers to cooperate, and, despite its seemingly chaotic development, the Internet acts as an efficient communication medium.

The functioning of the Internet also requires large-scale control of many complicated organizational issues and many important entities set the proper guidelines and policies. Examples can be found in the registration authorities that regulate the assignment of domain names and IP addresses,[11] or the handling of security-related problems. As we have shown, however, not one of the organizational or technical groups has ever plotted a global project of the Internet. The Internet is not driven by any supervising agent or authority, nor follows the blueprint of a pre-established architecture. It grows and develops because of cooperation and self-organization, to conform to technical standards and associative needs. Indeed, if we look at the Internet on a coarse grained scale, we see a spontaneously growing system, whose large-scale dynamics and structure are a cooperative effect due to many interacting units aimed at optimizing local communication efficiency.

---

[11] There are currently four Regional Internet Registries: the Réseaux IP Européens Network Coordination Centre (RIPE), the American Registry for Internet Numbers (ARIN), the Asia Pacific Network Information Centre (APNIC), and the Latin American and Caribbean IP address Regional Registry (LACNIC).

2

How the Internet works

While the scope of this book is to look at the Internet as a self-organizing complex system and to study its large-scale properties by using a statistical approach, a general knowledge of how the Internet works is necessary to identify the main elements forming the Internet, as well as their respective interactions. To give a physical analogy, if we want to understand the properties of a certain material we first have to know the atomic elements it is composed of, and how they interact. Similarly, it is impossible to approach the Internet without first having some hint of what a router is and how it communicates with its peers.

In this chapter we provide a brief description of the different elements, both hardware and software, at the basis of the Internet's functioning, to allow the non-expert readers to familiarize themselves with the general mechanisms that make it possible to transfer data from host to host. These mechanisms will turn out to be very relevant for understanding problems related with measurement infrastructures and other communication processes taking place in the Internet. Our exploration of the Internet's workings will be necessarily non-technical and, needless to say, the expert reader can freely skip this chapter and use it later on for convenient reference.

2.1 Physical description

In providing a general picture of the Internet, the starting point is the concept of a *computer network*. A network is any set of computers – usually referred to as *hosts* – connected in such a way that each one of them can inter-operate with all the others. The connection among hosts is made possible by two major components: hardware and software. The hardware refers to the physical components of the networks, such as computers and communication lines, ranging from the local telephone lines to fiber optic cables, and even satellite connections, that transfer data between computers. Software refers to the set of computer programs that rule the exchange of data through the hardware components. The software defining the network operations are often called *protocols*, since they define the set of standards that allows the handling of communications.

The Internet itself is a network of heterogeneous networks mutually interconnected. Among the various types of networks, we can identify two basic configurations, mainly determined by their geographical size:

- Local Area Networks (LANs). These kind of networks are used to connect hosts inside limited areas (buildings, university departments, etc.). They can use different technologies and protocols, such as the Ethernet, token rings, etc.
- Metropolitan Area Networks (MANs) and Wide Area Networks (WANs), on the other hand, connect computers which are scattered over wide geographical areas by using fiber optics cables, long distance land lines, radio, or satellite transmission.

The various networks composing the Internet are connected via specific devices called *routers*, that rule the communication among hosts in both the same and different networks. Routers play a key role in the Internet since they are not simple data forwarders but provide the physical connectivity of the Internet. They also continuously exchange updates on the network status, routing paths, and other vital information needed to keep alive the Internet, and choose the best routing for data.

There are other important components of the Internet, such as bridges, repeaters, switches and public exchange points. Bridges are devices that connect two or more networks by forwarding only a certain kind of traffic, i.e. they act as filters. Repeaters are devices that just propagate the signal from a cable to another, without taking any routing decisions. Switches are devices that join multiple hosts together in LANs.[1] While the previous components usually work at the LAN level, exchange points interconnect autonomously administered networks in public points where several routers are interconnected by the so-called shared media.

The large heterogeneity of the Internet is reflected also at the software level, since each network can rely on different protocols. Internet routers therefore have to be able to translate from one protocol to another, by working on the basis of a general protocol that plays the role of the universal language and addressing system of the Internet. The Internet is indeed built on a whole family of cooperative protocols often referred to as the Internet *protocol suite*.

---

[1] Switches are essentially high performance hubs for the local connection of hosts. Technically speaking, when a set of computers is connected to a hub and two of those computers communicate with each other, hubs simply pass through all network traffic to each of the two computers. Switches, instead, are capable of determining the destination of each individual traffic element, selectively forwarding data to the actual destination host.

2.2 Protocols

The Internet is a *packet switched* network. This implies that whoever is using the Internet does not have a dedicated piece of the network working exclusively for him. This philosophy is the opposite to *circuit switched* networks – such as, the telephone system – in which, when a call is operated, a section of the network is specifically assigned to establish a dedicated connection, or circuit, between two points. This implies that that specific part of the network is unavailable to all other users, even when the call is put on hold. In a *packet switched* network instead, all the communications between two hosts are mixed together with everyone else's data, put in common pipes, delivered to the specified destination address, and only there finally sorted out again. The entities supervising this process are Internet routers, which perform all these operations by following a set of standard protocols.

The Internet TCP/IP protocol suite[2] contains a family of protocols of which the *Transmission Control Protocol* (TCP) and the *Internet Protocol* (IP) are the most important ones. The IP defines a unique address space for the Internet in which each host receives its own IP number, also called IP address. When a host sends a packet of data to a given address, the router forwards it to the destination address. Necessarily, routers do not have a physical connection to all other routers in the Internet. The router handles the packet by looking at the destination address and sending it to the neighboring router[3] closer to the destination address, i.e. the best *next hop* toward the final destination. The way the router decides which is the next hop router is determined by the *routing protocol* algorithms, which shall be discussed in Section 2.4. This feature provides the network with the great advantage that all routers are equally important: The failure of any one of them does not preclude the network functioning, since routers can decide in real time to forward packets through a different path.

In a packet switched network large blocks of data are segmented into smaller parcels of similar size, which are sent separately. In this way all users equally share the resources, and, in case of heavy traffic, the performance becomes democratically worse for everybody. The main role of the TCP is to break the information down into packets of small and manageable size, stamped with the origin and destination IP.[4] Each data packet is numbered and labeled by the TCP so that

---

[2] There exist two main models of packet switched network in the Internet. One gets its name from its two main protocols and is referred to as the TCP/IP suite. A second one is the Open System Interconnection (OSI) suite. The latter has been developed by the International Organization for Standardization (ISO) on a conceptual model that defines seven layers: the application layer (the higher), the presentation layer, the session layer, the transport layer, the network layer, the data link layer, and the physical layer. In this framework the TCP and IP protocols work at the transport and network layer of the OSI protocol, respectively.

[3] The neighbors of a router are all the routers to which it is physically connected, i.e. the router's peers.

[4] For information about the statistics of IP packet sizes, see Claffy, Miller and Thompson (1998).

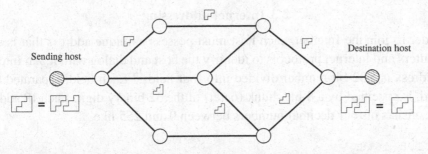

Fig. 2.1 Sketch of a packet switched network. The TCP breaks information from the sending host into packets that are transmitted by the routers (hollow circles) to the destination host, where the TCP reassembles them back into the original data.

the TCP software can collect and reassemble the data packets in the proper order on receipt by the destination host, see Figure 2.1. The TCP thus allows the packets to be sent through different paths or unsequentially, providing further flexibility to Internet data transmission. Things, though, are not always ideal on the Internet; packets can be lost or suffer large delays. Indeed, when a router gets overloaded, it starts to build up a queue of data packets to be handled. If the number of packets in the queue becomes excessively large – overflowing the memory buffer – the router just discards newly arrived packets. The TCP handles these situations by sending re-transmission messages and executing checks of the transmission delays that allow the sending host to re-send packets only if really necessary.

While ensuring that messages are correctly transmitted, the setting up of a TCP communication requires a lot of overhead in the amount of traveling packets and CPU time. For this reason, the Internet protocol suite also implements other standard transport protocols, especially convenient for programs that only send short messages. For instance, the User Datagram Protocol (UDP) does not take care of missing packets or keeping data in the right order. UDP just sends short messages without waiting for a reception answer. We shall see that light packets transport protocols, such as the UDP and the Internet Control Message Protocol (ICMP),[5] are generally used to send packets with the purpose of probing the Internet's structure.

The Internet protocol suite also contains other application-level protocols, many of them well known to Internet users. Among them we can name the File Transfer Protocol (FTP), the Telnet Protocol, and the Simple Mail Transfer Protocol (SMTP). These are higher-level software that control large file transfers between hosts, remote sessions on a given host, and e-mail exchange, respectively.

---

[5] This transport protocol is specifically devised as a tool for network maintenance and control.

2.3 Internet addressing

In order to join the Internet, each host must possess a unique address that is used by routers and Internet protocols to identify the host and dialog with it. An Internet IP address is a 32-bit number divided into four fields.[6] Each field, separated by a period, is specified by a 8-bit chunk (octet) of the 32 binary digit. Each IP address thus consists of four decimal numbers between 0 and 255 like

```
140.105.16.8,
197.12.33.128.
```

Since the Internet is a network of networks, each address is divided into a network portion and a local portion. The first part of the address (the network prefix) tells the routers what network the computer belongs to, while the last part specifies which host the address is referring to. IP addresses are divided into classes, depending on how many octets define the network portion. Commonly there are three classes of addresses:

Class A These addresses are assigned to very large networks with many hosts. Only the first octet is reserved for the network portion and the last three octets are for the local part. For this class the first octet ranges from 0 to 127. It must be noted that the zeros in the local portion are usually disregarded; for instance, the number 46.0.0.0 will be referred to as network 46.

Class B For this class the first two octets represent the network portion and the last two octets the local portion. This class is used for large networks, such as campuses or some WANs. The first octets range from 128 to 191, while the last octets are fully available in the range 1–254 (255 is a reserved number for global broadcasting). Class B networks can support in principle up to 65,532 hosts.

Class C These addresses are usually assigned to LANs. The first three octets are reserved for network identification and only the last octet for the local part. The first octet of class C networks is in the range 192–223 and allows in principle a limited number of hosts (up to 254).

In addition, there are two special network classes, D and E, reserved for Internet multicast and experimental use.

The network prefix is assigned by registration entities that keep registries of used numbers.[7] The local part of the number is assigned to the actual hosts by the network administrators. Individual networks can expand in different extensive LANs

---

[6] The IP address system described here, also known as IPv4 protocol, allows in principle up to $2^{32} - 1 = 4,294,967,295$ different addresses. The next generation of IP address (already implemented in some network applications), the IPv6 protocol designed by the IETF, works instead with 128-bit numbers, thus allowing a much larger number of hosts (Morton, 1997).

[7] The IP address space is distributed in a hierarchical way. The Internet Assigned Numbers Authority (IANA) allocates blocks of IP address space to Regional Internet Registries (RIRs). RIRs allocate blocks of IP address space to Local Internet Registries, who assign the addresses to the networks.

that, for routing convenience, should be considered as a single network showing a single address to the Internet. A solution to this demand is provided by the implementation of *subnetting*. Subnetting provides an additional level between the network and local portion of the IP address, in such a way that part of the octet of the local address is used to identify different subnets. At the practical level, this is done by means of a *net mask* that specifies the total number of bits identifying the network.

Finally, we want to introduce a different kind of network identification that refers to *autonomously administered networks*. Each autonomous system (AS) is a single administered entity, that may correspond to many routers and networks, which autonomously determines internal communication and routing policies. As a first approximation we can assign each AS to an Internet Service Provider (ISP). It is worth stressing that an AS may be geographically very delocalized. The increasing complexity of the Internet has led to the separation of the routing internal to each AS from the routing among ASs; i.e. intra-domain from inter-domain routing. Each AS has thus been assigned an identifying 16-bit number. AS numbers are used by the inter-domain routing to choose paths complying with service agreements among the various ASs as we shall see in the next section.

2.4 Routing packets

Routers deliver data packets by forwarding them to a peering router along the most convenient path from the origin to their final destination. In order to know what is the most convenient path, routers must have a rather global knowledge of the Internet. In order to manage and update this knowledge, routers continuously exchange information on the status and connectivity of the network. Information propagates from the router to its neighbors, in a hop-by-hop propagation process. Routing protocols rule this process, ensuring that all routers converge to a best estimate of the path leading to each destination address.

At present, Internet routing is split into two parts. The Interior Gateway Protocol (IGP) is used within LANs and single administered domains. The Exterior Gateway Protocol (EGP), alternatively, routes information among different networks. Both protocols provide great network stability by guaranteeing that if a connection is malfunctioning, the network can rapidly adapt to send data packets to the destination along a different path. Figure 2.2 provides an illustration of the global Internet routing architecture.

The standard IGP within networks is the Routing Information Protocol (RIP). It is based on three main elements: (i) a database that stores information on the shortest route from computer to computer; (ii) a propagation mechanism for routers to exchange their databases; (iii) an updating algorithm that allows the router to

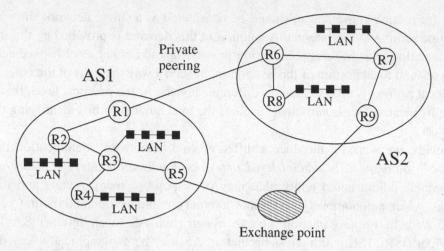

Fig. 2.2 In the Internet, autonomous systems are interconnected by a private peering relation (routers R1 and R6) or public exchange points (routers R5 and R9). Routers R1, R5, R6, and R9 use an exterior gateway protocol such as the BGP, while the remaining routers adopt an IGP only aimed at internal routes.

update its database when a shorter path is announced by its neighboring routers. The database stores, for every address in the same RIP network, the address prefix, the neighboring routers to which to send a message addressed to IP addresses within that prefix, and the number of routers between the given router and the one that can ultimately deliver the message directly to its destination. In addition, a flag is added to all the information that has recently changed in the database, so that communicating routers can update their own databases. At regular intervals each router sends an update message containing its routing database to the routers directly connected to it. When a router finds that a neighbor has a shorter path to any other router, it will update its own database following a pre-established mathematical algorithm and, in turn, it will communicate the update to its neighboring routers. In this way the update is quickly propagated to the whole network, allowing the adaptation to rapid changes in connectivity due to outages or the appearance/disappearance of computers.

If a data packet is sent outside the RIP network, it is directed to the router responsible for the EGP. This router will send it across the Internet to the EGP router of the destination address network, that will forward the packet to its final destination using the local RIP. Beyond the simple shortest path selection, however, inter-domain routing has to consider policy and commercial agreements among ISPs. For economy reasons, the routing paths in the EGP are specified in terms of AS numbers, avoiding in this way the use of long lists of intermediate routers.

```
*> 192.9.9.0    134.24.127.3     1740 701 90
*               204.212.44.128   234 266 237 3561 701 90
*               205.238.48.3     2914 1 90
*               144.228.240.93   1239 701 90
*               204.70.4.89      3561 1 90
*               194.68.130.254   5459 5413 1 90
*               202.232.1.8      2497 701 90
*               158.43.133.48    1849 702 701 90
*               131.103.20.49    1225 2548 1 90
```

Fig. 2.3   Example of a BGP routing table from http://moat.nlanr.net/ ASx/.

The most used EGP is the Border Gateway Protocol (BGP). When a BGP router is turned on, it establishes a *peering* session[8] with the directly connected BGP routers and downloads their entire routing tables. After that operation, all other information exchanges are just simple update messages. Update messages contain the withdrawal of certain routes or a new preferred route for a certain network prefix. In the latter case, the BGP router announces to all its peers the reachability of a particular prefix, through a certain path traversing one of its neighbor BGP routers after prepending its own AS number to the AS path. If the update contains new information, the receiving BGP routers run a decision algorithm to decide which peering router has the best routing path for the prefix address specified in the update. The algorithm for the decision process contains a *cost* field that allows to take into account specific policy criteria determined by the local administrator, considering in particular the mutual agreements reached among ISPs. Finally, if the new path is adopted, the receiving router will announce in its turn the new path to its peers.

This nearest neighbor spreading procedure finally converges, and BGP routing tables appear as collections of entries, such as the one shown in Figure 2.3. In this example, the BGP router can reach the address space 192.9.9.0 through nine different paths. In the first column the address of the next hop router is specified. The remaining columns lists the ASs in the path which are traversed to reach the destination. For each address prefix the best path is marked with a '>' sign. Alternative paths can be used in the case of outages or other circumstances. We shall see in Chapter 3 that BGP tables are extremely important to derive a map of Internet connectivity at the AS level.

Other important information exchanged among BGP routers is the *keepalive* message. Each router sends to its peers a keepalive message every 3–30 seconds. If a larger time elapses between two consecutive keepalive messages, the neighboring

---

[8] Peering refers to physically connected routers that agree to exchange traffic and routing information.

routers close the peering session and proceed to announce the change to the whole network. This allows the rapid handling of transient network damages.

2.5 The domain name system

So far we have depicted a rather complex address and routing system. While it is not crucial for the rest of this book, we want to mention here what makes the Internet easier to navigate by the common users. In fact, while numbers are perfectly fine for the communications among machines, it is not easy for humans to handle IP addresses. For this reason, computers on the Internet are given names, that are in a one-to-one correspondence with their IP address.

As it is easy to imagine, the naming of IP addresses introduces its own problems. The name administration is handled by a system called domain name system (DNS), which gives to different groups the responsibility for various subsets of the names. Each level in this system is called a *domain*, and they are separated by periods. For instance, the name `lyre.th.u-psud.fr` corresponds to a specific host with a precise IP address. The name `lyre` is assigned by the theory group (`th`), member of the University of Paris-Sud (`u-psud`), that belongs to the French national domain (`fr`). Each group is responsible for assigning the names at the lower levels. For instance, the `u-psud` group can create another group called `smt.u-psud.fr` because they need to name a new laboratory. This hierarchical responsibility in giving names ensures the uniqueness of the correspondence between IP address and DNS name. When routers need to translate names in IP numbers they ask some dedicated machines, called name servers, that look at their own list of correspondences. If the name server does not find the named address in its database, it asks the root server, that is a name server of the higher level zone, and so on up and down the hierarchical levels until the necessary information is retrieved.

It is important to stress that domains do not necessarily have a direct correspondence with single administered domains or networks, and that computers may have multiple names. Most importantly, the DNS is not necessary for communication and should be just considered as something that helps the Internet to be a comfortable place to work in.

3

Measuring the global Internet

The characterization of how routers, computers, and physical links interconnect with each other in the global Internet is a very difficult task due to several key features of network development. A first one is the Internet's size and continued growth. The Internet is growing exponentially and its size has already increased by five orders of magnitude since its birth. In other words, the Internet is a large-scale object whose global properties cannot be inferred, in general, from local ones. A second difficulty is the intrinsic *heterogeneity* of the Internet, that is it is composed of networks engineered with large technical and administrative diversity. The different networking technologies are merged together by the TCP/IP architecture that, while providing connectivity, does not imply uniform behavior. Moreover, networks range from small local campuses to large transcontinental backbone providers. This difference in size is reflected in different administrative policies that make routing through the Internet a highly unpredictable and heterogeneous phenomenon. Also very important is the fact that the Internet is a *self-organizing* system, whose properties cannot be traced back to any blueprint or chart. It evolves and drastically changes over time according to evolutionary principles dictated by the interplay between cooperation (the network has to work efficiently) and competition (providers wish to earn money).[1] This means that routers and links are added by competing entities according to local economic and technical constraints, leading to a very intricate physical structure that does not comply with any globally optimized plan.

The combination of all these factors result in a general lack of understanding about the large-scale topological structure and performance properties of the Internet. In its turn, this poor knowledge of the Internet is also starting to affect providers, who do not have the tools available to evaluate and forecast growth

---

[1] This also implies that, in general, Internet providers are not willing to share topological information on the networks they administer.

trends and performance problems. For these reasons, in recent years, several research groups have started to deploy technologies and infrastructures in order to obtain a more global picture of the Internet. Several studies aimed at tracking and visualizing Internet large-scale topology and/or performance are now providing Internet mapping projects at different resolution scales. These projects typically collect data on Internet nodes (routers, domains, etc.) and links in order to create a graph-like representation of large parts of the Internet. Along with these topological measurements, performance analyses are carried out by measuring transmission times among hosts as a function of the path, traffic load, or other parameters.

In the present chapter we focus on measurements of the global structure and performance of the Internet, and the corresponding graph-like representation. We will describe several measurement projects at different resolution levels (Murray and Claffy, 2001), with particular emphasis on the different probes and methods used to obtain information on Internet topology, the data they actually provide, and their positive and negative sides. The datasets obtained from these measurements will be at the core of the statistical analysis of the topological properties presented in Chapter 4.

3.1 Mapping of the Internet

As we have seen in the previous chapters, routers are the basic elements for Internet traffic. Therefore, the Internet is usually viewed as an undirected graph,[2] in which vertices (nodes) represent routers and edges (links) represent the physical connections between them. This picture does not model individual hosts, as they are too numerous, and neglects the characteristics of the links, such as bandwidth, actual traffic load, and geographical distance. For these reasons, the graph-like representation must be considered as an overlay of the basic topological structure, the *skeleton* of the Internet.

Despite the fact that the Internet grows without any central administration, an underlying hierarchy can be identified, for administrative and technical reasons. Today the Internet can be partitioned into autonomously administered *domains*, called autonomous systems (AS), which vary in size, geographical extent, and function. Each AS may exercise traffic restrictions or preferences, and handle internal traffic according to different autonomous policies. In general these autonomous regions can be classified either as *transit* or *stub* ASs. Transit ASs correspond to large backbones providing national or inter-continental connectivity, or to regional providers serving metropolitan areas. Stub ASs generally corresponds to campus networks, small local dial-up providers, or other collections of local area

---

[2] See Appendix A1 for an introduction to graph theory.

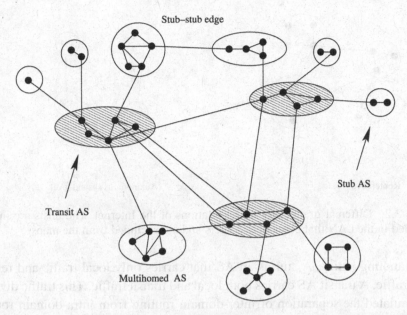

Fig. 3.1 The Internet domain structure. Filled circles represents individual routers. Hollow regions and shaded regions correspond to stub and transit ASs, respectively. Figure adapted from Zegura *et al.* (1997).

networks (LANs). The main purpose of transit ASs is to provide connectivity to stub ASs without having stubs directly connected to each other. For this reason, routers in transit regions are well interconnected and link a number of stub ASs at their *gateway* nodes. We can therefore topologically depict the Internet as a core of interconnected backbones to which regional, campus, and corporate networks are linked. Figure 3.1 represents a scheme of the Internet's domain structure.[3] This picture, however, is a very schematic approximation. For instance, multi-homed stub ASs connect to more than one transit AS. As well, stub-to-stub edges and transit ASs that only connect to other transit ASs are also present in the Internet. All these possibilities have led to more complicated hierachical characterizations, usually based on the level of connectivity (see Section 4.5).

A basic characteristic of the domain hierarchy is that traffic paths between nodes in the same domain stay entirely within that domain. For instance, stub ASs just handle traffic that originates and terminates inside the AS borders, while a routing path between two nodes in different stub ASs goes through one or more transit AS.[4] This introduces a more precise identification of AS classes based on their

---

[3] It is worth mentioning that the early Internet was quite different, with an essentially tree-like inter-domain hierarchy. The NSFNET provided long-distance connectivity to regional providers. The latter in their turn were offering connectivity to campuses and other LANs.

[4] Exceptions to this general picture can be found in the case of stubs connected by a stub-to-stub link.

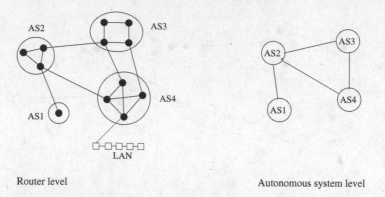

Fig. 3.2   Different granularity representations of the Internet. The hosts are included in the LAN that connect to routers and are excluded from the maps.

traffic handling policy. A stub is an AS that carries only local traffic and refuses
transit traffic. A transit AS carries both local and transit traffic. This traffic division
has stimulated the separation of inter-domain routing from intra-domain routing
and the introduction of the AS numbers. Each AS is identified in the Internet by
a 16-bit AS number used for routing purposes by the Exterior Gateway Protocol
(see Section 2.4). ASs were originally only a routing concept, but have eventually
led to a more general abstract representation of the Internet. Also in this case we
can use a graph-like representation to study the inter-AS connectivity, in which
vertices represent ASs and edges are *peering* relationships[5] among them.

Mapping projects focus essentially on two levels of topological description.
First, by inferring router adjacencies it has been possible to measure the Internet Router (IR) level graph. The second mapping effort concerns the AS level
graph of the Internet obtained from AS routing path information. Although these
two graph representations are related, it is clear that they describe the Internet at
rather different levels (see Figure 3.2). In fact, the collection of ASs and the inter-
domain routing system defines a coarse-grained picture of the Internet in which
each AS groups many routers together, and links are the aggregation of all the
individual connections between the routers of the corresponding ASs. This difference in scale implies that the corresponding graphs at the IR and AS levels can
show quantitatively different properties, although we expect them to show similar
large-scale statistical behavior. Data collection techniques for the two levels are
also different and both representations may be incomplete or partial to different
degrees. In particular, measurements may not capture all the nodes present in the
actual network and, more often, they do not include all the links among nodes. It

---

[5] AS *peering* is the relationship between two ASs that agree to exchange traffic and routing information through
one or more directly connected routers.

is therefore important to review the various data collection projects and discuss their limitations and their reliability when obtaining realistic representations of the Internet's structure.

3.2 A Ptolemaic view

In order to obtain Internet connectivity information we can inspect routing tables and paths stored in each router (passive measurements) or directly ask the network with a software probe (active measurements). In the latter case we can send IP packets on the Internet that elicit a reply from the targeted host. By collecting the information on the packets' path to the various destinations it is possible to build the graph of router adjacencies. This kind of probe has long been used by system administrators and is embodied in the `traceroute` tool. The `traceroute` command sends out hop-limited IP packets toward a given destination. Each packet is assigned a specific time-to-live (TTL) that is decreased by one at each hop. If the TTL becomes zero, the network kills the packet and forwards a death message to the sender. By incrementing the TTL by one at each successive trial, it is possible to progressively explore intermediate IP hops in the forward path to any given address, while avoiding the chance of packets wandering around indefinitely. Each hop in a `traceroute` path thus corresponds to a router that echoed the packet by sending back its IP address. Collecting all the response IP addresses allows the determination of the path the packet followed to reach the target machine. Figure 3.3 shows a typical output from the `traceroute` command.

Mapping efforts are generally based on computing router adjacencies from `traceroute`-like sequences sent to a list of networks in the Internet. The simplest choice is to trace paths to a list of destinations from a single network node. Early projects were using databases with 10^3–10^4 different hosts (Pansiot and

```
sinera:~> /usr/sbin/traceroute lyre.th.u-psud.fr
traceroute to lyre.th.u-psud.fr (129.175.117.133), 30 hops max, 38 byte packets
 1  * * *
 2  phc3 (147.83.54.1) 1.603 ms 1.237 ms 1.144 ms
 3  sistole-routing (147.83.124.125) 1.612 ms 1.467 ms 1.676 ms
 4  10.10.124.30 (10.10.124.30) 1.838 ms 3.247 ms 2.981 ms
 5  10.10.124.14 (10.10.124.14) 2.659 ms 2.700 ms 1.763 ms
 6  AT0-2-0-0.EB-Barcelona0.red.rediris.es (130.206.202.77) 3.493 ms 5.022 ms 4.854 ms
 7  AT6-2-1-0.EB-IRIS4.red.rediris.es (130.206.224.1) 15.606 ms 15.733 ms 13.913 ms
 8  GE1-0-0.EB-IRIS1.red.rediris.es (130.206.220.25) 17.345 ms 15.039 ms 22.508 ms
 9  rediris.es1.es.geant.net (62.40.103.45) 21.434 ms 22.056 ms 26.842 ms
10  es.fr1.fr.geant.net (62.40.96.70) 52.637 ms 57.271 ms 60.588 ms
11  renater-gw.fr1.fr.geant.net (62.40.103.54) 62.144 ms 58.640 ms 54.860 ms
12  orsay-a0-0-500.cssi.renater.fr (193.51.179.158) 53.077 ms 54.794 ms 52.920 ms
13  univ-paris-sud-orsay.cssi.renater.fr (193.51.183.29) 49.734 ms 48.302 ms 46.284 ms
14  129.175.127.35 (129.175.127.35) 41.557 ms 41.629 ms 33.746 ms
```

Fig. 3.3   IP sequence output of a `traceroute` command from the host `sinera.upc.es` to the host `lyre.th.u-psud.fr`.

Grad, 1998), while recent extensive measurements are sending IP packets to about 10^5–10^6 registered networks (Burch and Cheswick, 1999; Govindan and Tangmunarunkit, 2000; Huffaker, Plummer, Moore and Claffy, 2002). With such large scanning efforts, the `traceroute` technique is cast into a customized program that selects target hosts and executes many traces in parallel. Packets are usually sent to public access sites on the selected network or to a valid address guessed by the program.[6]

An important issue in mapping projects is that the used packet probes must not be confused with hacking attempts. For this reason, paths are usually discovered one hop at a time by progressively increasing the packet TTL, and tracing attempts are repeated over long time intervals. Once the packet reaches the host, the path is stored and the target address is saved for routinely repeating the tracing process. In fact, the path may very well vary between different traces at different times, because of momentary outages, congestions, network reconfigurations, or routing policy changes. The tracing program thus executes a full scan in a time window that varies from one day for intranetworks to weeks for large sections of the Internet. The collection of all the paths finally provides a picture of the physical Internet connectivity that, in the case of intranets, can be refined up to the host level.

It must be clear, however, that tracing routes from a single source does not provide a complete map of the Internet. At least one path from the source to each node of the graph is found, but the interconnections among the addressed nodes are not necessarily discovered, see Figure 3.4. In other words, in a shortest path routed network, one might expect that path probing results in a tree structure rooted at the source host, missing all the cross-links among intermediate nodes in the paths (sometimes referred to as *lateral connectivity*). Technically speaking the resulting map is a collection of *spanning trees* to the targeted addresses. This amounts to a "Ptolemaic" point of view, in which the Internet map is centered at the probing source; a very partial and source-based view of the Internet world indeed. This problem is mitigated by two technical facts that help path probing to actually discover a richer connectivity structure (see Figure 3.4). First, inter-domain routing is policy based, and it may imply that different paths cross the same IP addresses. In addition, the probing is routinely repeated on time windows spanning several days, thus potentially discovering *backup paths* to several target addresses (Govindan and Tangmunarunkit, 2000). Despite these "bonuses," however, it is clear that mapping the Internet from a single location misses a large number of cross-links in the Internet map.

The partial discovery of cross-connectivity is not the only problem related to the measurement of the Internet's router structure when using active probing

---

[6] For instance, the network `140.105.16.0` very likely has a host named `140.105.16.1` on it.

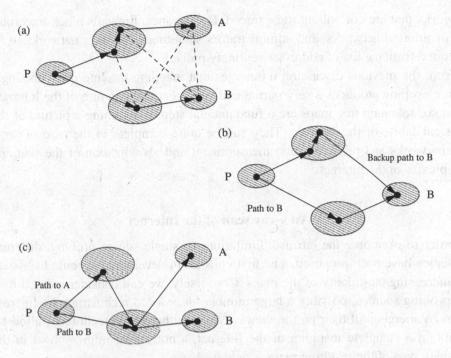

Fig. 3.4   (a) Single source spanning tree from the probing host *P* to destinations *A* and *B*. Filled circles correspond to routers and shaded areas represent ASs. The dashed edges represent non-detected router adjacencies. (b) Backup paths are discovered in repeated measurements, providing supplementary topological information. (c) Routing policies can lead to a more complete view by imposing constraints on the possible paths. Figure adapted from Govindan and Tangmunarunkit (2000).

strategies (Burch, 1999; Govindan and Tangmunarunkit, 2000). Routers have by definition several *interfaces*, each one corresponding to a different IP address. We therefore have multiple *aliases* for the same router. Yet, the different IP addresses of a router's interfaces reported by the `traceroute` are assigned to different nodes on the map, despite belonging to a single physical entity. This creates a false connectivity that might introduce spurious results in the topological description of the graph. A detailed example of this effect is discussed in Appendix A2. As well, some IP addresses are shared by more than one machine. This should not happen, but sometimes it does, resulting in different physical routers being grouped in a single node. This is the reason why the original maps collected by path-probing methods can be more precisely considered as IP connectivity graphs. A further problem is related to the choice of the addressable IP space; i.e. which prefix might contain addressable nodes. While these disadvantages, as we shall see in the following, can be minimized by developing some specific tools, a final and more knotty problem concerns security measures and the existence of networks that are not willing to be traced. For instance, firewalls block tracerouting of internal networks, and administrators sometimes ask their networks to be excluded from the list of addresses routinely probed.

From the previous discussion it emerges that mapping the Internet with single source probing produces a very partial and "ego-centered" picture of the Internet. Even so, spanning tree maps are a fundamental step in obtaining a picture of the physical fabric of the Internet. They can be quite complete in the case of large intranetworks and provide a first measurement and visualization of the size and complexity of the Internet.[7]

3.3 An x-ray scan of the Internet

In order to overcome the intrinsic limitations of single source probing, different strategies have been proposed. The first consists in developing specific heuristics for increasing the fidelity of the maps. Conversely, we can consider using different probing sources, possibly a large number, in order to reconstruct the Internet maps by merging all their partial views. While neither of the techniques allow us to obtain a complete mapping of the Internet, a noticeable improvement in the completeness of the resulting maps is achieved.

3.3.1 The heuristic approach

One of the largest efforts in using heuristics to improve Internet scanning is represented by the *Mercator* Internet mapping program (Govindan and Tangmunarunkit, 2000). To discover the Internet map, *Mercator* works by routinely selecting a prefix from its population of IP networks and probing the path to an address selected within that prefix with a `traceroute`-like tool.[8] Additionally, however, several heuristics are implemented to derive IP targets, discover cross-links, and resolve router interfaces.

Since *Mercator* is intended to map the Internet with nearly zero initial information, it does not use any external databases to drive the map discovery. While a possible choice in this case could be to probe paths to randomly chosen addresses from the entire IP address space, *Mercator* uses instead a heuristic defined as *informed random address probing*. The goal of this method is to infer which portion of the IP address space actually contains addressable nodes. The informed random address probing starts from a seed prefix (usually the IP address of the source host),

---

[7] See for instance the Internet Mapping Project at Lumeta Corporation, http://research.lumeta.com/ches/map/index.html.

[8] *Mercator* sends UDP packets with increasing TTLs, and termination conditions if a loop is detected in the path or the probe fails to elicit a response.

and increases its prefix population along with the mapping process. In practice, whenever a path to a responding address *A* is found, *Mercator* assumes that the prefix of *A* must contain addressable nodes and then expects that also neighboring prefixes will likely contain addressable nodes.[9] Thus, if in a given time window the prefix population has not increased, *Mercator* selects a neighbor of an existing prefix, chosen with a probability proportional to the number of successful probes addressed to that prefix (a probe is considered successful if it discovers at least one previously unknown router). The same preferential mechanism is applied to the choice of prefixes selected for path probing. This feature biases the prefix discovery and the mapping process towards prefixes densely populated with addressable nodes.

Another important heuristic that *Mercator* implements is aimed at the discovery of "cross-links." As discussed in the previous section, single source probing misses a conspicuous amount of cross-links which are not discovered in the tree-like exploration of paths. In order to increase the likelihood of discovering links in the Internet graph, *Mercator* makes use of another feature called *source-routed path probing*. In simple words, the software directs path probes to its prefix population from detected *source-route* capable routers. These routers allow the use of path probing as if they were the source of the packets and are somehow equivalent to the acquisition of additional probes that scan the network from different viewpoints. This different perspective allows the discovery of alternative paths to already targeted address, see Figure 3.5. In principle, each source-route capable router could provide the same amount of data as if *Mercator* were running at that location. Unfortunately, source-route capable routers are very few, and their locations cannot be opportunely chosen. In addition they cannot be intensively used, since a large probing overhead usually triggers the system administrator alarm,

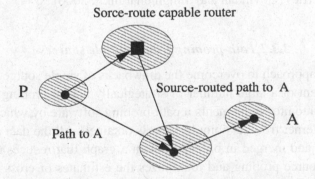

Fig. 3.5   Source-routed path probing allows the discovery of cross-connectivity by providing a different perspective for probing. The source-route capable routers is marked as a filled square.

---

[9] This technique is based on the assumption that address registries allocate the address space sequentially.

which does not always imply a cooperative reaction. How much the use of these routers is helping to map Internet connectivity is therefore related to several factors, including their number and location with respect to the original probing host, which cannot be exactly quantified. Nevertheless, numerical experiments show that, in sparse graphs, even a rather small fraction of source-route capable routers can improve considerably the amount of discovered adjacencies (Govindan and Tangmunarunkit, 2000; Vázquez, Pastor-Satorras, and Vespignani, 2003).

Finally, a last heuristic of *Mercator* is devoted to routers' *aliases resolution*. As discussed in the previous section, path probes discover router interfaces, thus a router level map requires identification of all the interfaces (IP numbers) belonging to the same router. A simple analysis of the IP numbers does not allow the aliases of a given router to be distinguished. To solve this problem, *Mercator* sends alias probes based on a common feature of IP implementations. In simple terms, the IP packet is addressed to an unused port of the router interface. An error message is then sent to the source address, from an eligible interface of the router. If the interface is different from the one to which the original packet was addressed, then the two interfaces belong to the same router. In addition, and in order to maximize the number of resolved aliases, other refinements, such as repeating alias probes in time and emitting alias probes also from source-route routers, have also been implemented (Govindan and Tangmunarunkit, 2000).

Mercator has been used to collect one of the largest publicly available maps at the IR level. As previously stressed there are many "caveats" indicating that the resulting map could be still largely incomplete. However, validating tests on small and medium-sized ISP networks have proven to be rather successful, and there are reasons to believe that maps generated with well-developed heuristics could provide realistic topologies for protocol simulations and statistical studies of the Internet's properties (Govindan and Tangmunarunkit, 2000).

3.3.2 Path-probing from multiple sources

An orthogonal approach to overcome the drawbacks of single source path probing is the deployment of a large number of strategically placed probing monitors. In principle, each monitor implements a path-probing software by which a spanning graph of the Internet from its point of view is obtained. All the data are then centrally collected and merged in order to obtain a graph that reduces the problems due to single source probing and maximizes the estimates of cross-connectivity. In view of the evident analogies, this mapping effort has been dubbed *Internet tomography* (Claffy, Monk, and McRobb, 1999).

A large-scale project based on this philosophy is the Macroscopic Topology Project developed at the Cooperative Association for Internet Data Analysis

(CAIDA). Since 1998, this project has been performing measurements and developing visualization tools to collect and analyze large-scale Internet topology and performance. The primary measurement tool is represented by *skitter* that, similarly to other customized probing programs, sends increasing TTL traceroutes to collect forward paths from à monitor to addresses in a destination list. The project consists of several monitors divided into groups which probe different destination lists according to the different specific problems analyzed[10] (Huffaker *et al.*, 2002). Each monitor stores the obtained data in files which are transferred to a central repository. In turn the repository collector machine provides the destination lists for each monitor. In order to maximize the number of detected links, *skitter* monitors are placed strategically around the whole Internet and destinations are probed with different cycle times depending on the monitors' locations.

The IP graphs collected by *skitter* monitors reflect a significative part of the addressable space, and can be studied as a representative granularity of the Internet (Huffaker *et al.*, 2002; Broido and Claffy, 2001). However, since the goal is the construction of an IR level map, interfaces with different IP addresses must be identified with routers. Aliases resolution is crucial also in this framework since it could happen that different monitors probe the same router on different interfaces. To avoid this problem, CAIDA developed a tool called *iffinder* that is able to resolve aliases in a similar way as described previously for the *Mercator* program (Huffaker *et al.*, 2002).

CAIDA is producing an enormous amount of data sets that can also be used to highlight the role of specific backbones, traffic exchange points, and traffic bottlenecks. *Skitter* monitors probe the largest destination list available at the moment, and data have been collected over several years forming a database ready for the study of the Internet's dynamical evolution.

3.3.3 Large graph visualization

One of the problems faced in the analysis of Internet maps at the very large-scale level consists in their visualization. While there are several methods for looking at network data sets, the logical layout based on connectivity shown in Figure 3.6 is the most frequently used. Indeed, the first concern in a hypothetical Internet navigation is the number of intermediate router hops, more than the effective mileage between hosts. In addition, colors may be used to display link usage, ownerships, and time changes. This may help to find anomalies, such as bottlenecks or failure points.

---

[10] In October 2002, the 21 monitors probe approximately 9×10^5 destinations spread over about 70% of the globally routable network prefixes.

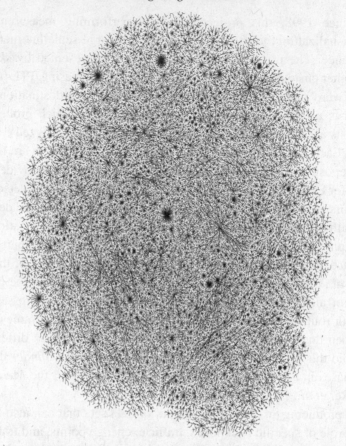

Fig. 3.6   Two-dimensional image of a router level Internet map collected by H. Burch and B. Cheswick, courtesy of Lumeta Corp. (http://www.lumeta.com).

Yet, the drawing of the macroscopic Internet structure is not an easy task, since the resulting graph datasets are huge and push to the limit the CPU power and resolution of current computers. For this reason, mapping projects allocate a large amount of resources in developing specific visualization tools. Despite these efforts, however, very large graph drawings, while visually appealing, are so dense and intricate that they elude any attempt to find an intelligible organization at the root of the seemingly disordered drawing of edges and vertices. This has stimulated Internet representations at the coarser AS level, and eventually has led to a more quantitative characterization based on statistical methods.

3.4 AS level maps

A coarser view of the Internet can be obtained by aggregating IP addresses or routers into their corresponding ASs[11] (see Figure 3.7). In this way CAIDA is

---

[11] Conversely, it is also possible to explore and visualize the connectivity of individual ASs with tools such as *Hermes*, http://www.dia.uniroma3.it/~hermes/.

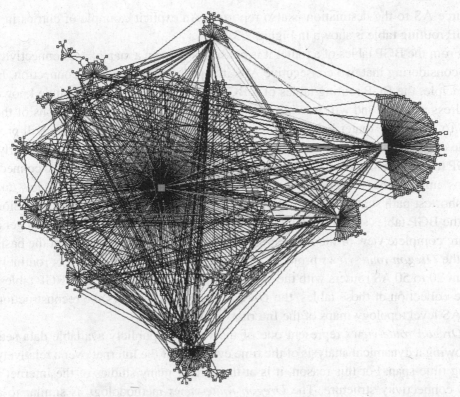

Fig. 3.7 Two-dimensional image depicting the Internet's AS connectivity reconstructed from *skitter* traces. After Claffy *et al.* (1999).

providing a visualization of the AS core by converting each IP address in the AS responsible for its routing. The mapping is made by using the BGP routing tables collected by the *Oregon route-views* project. BGP tables contain the AS paths to destination IP addresses, the AS at the end of each path being the administrative responsible for the corresponding address. This data aggregation allows AS connectivity maps to be reconstructed from active measurements and provides logical layouts that can be used to study the role of specific ASs in routing traffic across the Internet.

Interestingly, AS level graphs of the Internet can be obtained also by passive measurements.[12] In practice, AS graphs are reconstructed by collecting BGP routing tables from the routers in charge of the inter-domain protocol. BGP is a distance vector protocol that constructs paths by successively propagating path updates between peering BGP routers. This process produces BGP routing tables containing the list of all destination prefixes accessible from each BGP router. For each destination prefix, the path vector – listing the traversed ASs – from the

---

[12] A comparison between passive and active measurements, from the point of view of their respective accuracy and reliability, is discussed by Broido and Claffy (2001).

source AS to the destination is also reported. An explicit example of entries in a BGP routing table is shown in Figure 2.3.

From the BGP tables of a router it is possible to obtain a map of AS connectivity by considering that two consecutive ASs in a path define a peering connection. In principle, the BGP routing tables of a BGP router should cover the whole known address space,[13] and will provide a complete view of the peering relations of the AS to which the router belongs. Very likely, however, these BGP tables will pro-' vide only partial information on the interconnectivity among other ASs. Similarly, BGP tables obtained from a different AS will see only some, but not all the connections among other ASs, because each BGP router has its own particular view due to shortest path strategies and policy reasons. Therefore by considering the union of the BGP tables stored in many routers belonging to different ASs, we can get a more complete view of the AS interconnectivity. This consideration is at the basis of the *Oregon route-views* project,[14] where a dedicated router connects routinely from 20 to 50 AS routers with the specific purpose of collecting their BGP tables. The collection of these tables, the *Oregon route-views*, allows the reconstruction of AS level topology maps of the Internet.[15]

Oregon route-views represent one of the very few publicly available data sets allowing a dynamical analysis of the time evolution of the Internet over a relatively long time span. For this reason, it is at the core of many studies of the Internet's AS connectivity structure. The *Oregon route-views* methodology is similar to a multiple source probing of the Internet, where the probing paths of each BGP monitor are just its own BGP paths. *Oregon route-views*, thus, may not reveal many of the paths which are less advertised or not considered by the AS routers peering with the Oregon server. As in active path probing, by considering larger numbers of BGP tables these problems can certainly be minimized, but it is not possible to evaluate the completeness of the map obtained from the union of many BGP views. In other words, it is not known a priori how many BGP tables from different AS routers are needed in order to obtain a complete view of the Internet's AS interconnectivity.

With the aim of establishing the completeness of the AS level topology captured by the *Oregon route-views*, Qian *et al.* (2002) supplemented and compared these data sets with BGP summary information from a number of different sources. In

---

[13] This is not always the case. In particular when a BGP router does not have an AS path to a certain address, it forwards all traffic for that address to another default router. BGP routers that for any particular reason do not address a large section of the address space can be identified and not considered in the statistical analysis (Qian, Chang, Govindan, Jamin, Shenker, and Willinger, 2002).

[14] The University of Oregon's Route Views project, http://www.antc.uoregon.edu/route-views/.

[15] A daily archive of the *Oregon route-views* is made available from November 1997 to March 2001 at http://moat.nlanr.net/Routing/rawdata/ by the National Laboratory for Applied Network Research (NLANR) (McGregor, H-W.Braun, and Brown, 2000). From April 2001 raw data are available at the route-views archive http://archive.routeviews.org.

particular, several ISPs residing in different ASs allow public access to their BGP tables. Supplementary routing policy information can also be obtained from the Internet Routing Registry (IRR). This new information can be aggregated to the *Oregon route-views* to provide an *extended* AS graph of the Internet, to be compared with the original one. The extended maps contain 20–50% more physical connections (edges), but only 2% more ASs (vertices). This discovery demonstrates that the graphs obtained from *Oregon route-views*, missing a noticeable fraction of the Internet connectivity, might be rather incomplete. While this finding is extremely important for specific local studies, Internet graphs derived from *Oregon route-views* might still provide a reasonable statistical sampling of AS physical connectivity, sufficient to capture the correct large-scale topological and statistical properties of the Internet. Indeed, even large quantitative differences do not necessarily imply different qualitative features at the large-scale statistical level (see Chapter 4). The *extended* maps therefore become an essential benchmark from which to test the stability and consistency of the statistical properties of the original *Oregon route-views* maps in the case of larger connectivity samplings.

3.5 Internet geography

Until now we have only considered Internet maps as graphs representing the physical connectivity, and lacking any real distance or location information. This *topological* layout is in opposition to geographical layouts in which Internet elements are visualized through physical maps. The idea is therefore to find the geographical location of each vertex (router or AS), place the vertex at that very position, and draw lines between physically connected vertices. This strategy sounds simple enough but, unfortunately, determining the geographic location of a given IP address is a non-trivial task. The problem in finding a router's location resides in the fact that, due to business and security reasons, many ISPs do not want the exact positions of their machines to be publicly available. In many cases it is difficult to extract even an approximate location from the host name, and many routers just have an IP address.

Strategies to establish a correspondence between IP addresses, domain names, and ASs depend on the *whois* database, which provides the registered headquarter's address of ISPs. This procedure can of course generate errors, since large ISPs have sometimes routers scattered all over the world and not just at the headquarter's geographical location. Specific tools[16] therefore combine different heuristics and databases in order to minimize these problems.

---

[16] See for instance the NetGeo tool developed at CAIDA, http://www.caida.org/tools/utilities/netgeo/.

Besides map layout, geographic location represents very valuable information in all studies that correlate Internet connectivity and performance with the actual physical distance among routers (Yook, Jeong, and Barabási, 2002; Lakhina, Byers, Crovella, and Matta, 2002). These correlations will turn out to be extremely relevant when obtaining more realistic models of Internet graphs, as we will see in Chapter 5.

3.6 The Internet's global performance

Along with the study of topology, a large percentage of the activity in Internet measurement infrastructures is devoted to the analysis of workloads and performance. These are critical issues for assessing the overall health status of the network, since they are a quantitative measure of the efficiency and stability of its communication capabilities.

Traffic measurements focus on the number of IP packets generated by different kind of traffic (TCP, WWW, etc.) on specific Internet links. Usually data constitute the time series of traffic load at different resolution scale and in different time windows. This kind of measurements have been the first to be performed regularly to monitor the network behavior and have led to a large activity in the field of traffic signal (Willinger, Taqqu, and Erramilli, 1996). Recently, measurements have been extended to larger scales and data have been analyzed on the basis of source and destination addresses, providing a first attempt to correlate traffic flows with the network topology (Claffy, 1999; Huffaker, Fomenkov, Moore, Nemeth, and Claffy, 2000; Uhlig and Bonaventure, 2001).

The basic testing tool for evaluating Internet performance is the original `ping` (Packet InterNet Groper) program. Based on the Internet Control Message Protocol (ICMP), `ping` sends packets that elicit a reply from the target host. The program then measures the round-trip-time (RTT), i.e. how long it takes for each packet to make the round trip to its destination. Other features, such as the measure of the number of packets lost in a certain time window, are also implemented. The NLANR and CAIDA measurement infrastructures use `ping`-like probes from their monitors to collect RTT and packet loss data for hundreds or thousands of Internet destinations. Other organizations and projects, such as the PingER monitoring infrastructure[17] and the Réseaux IP Européens (RIPE),[18] have also a number of sites sending regularly ICMP probes to a few hundred targets to monitor the end-to-end performance of Internet links.

RTT data quantify the speed at which IP packets travel across a specific path. By comparing RTTs on different paths and at different times it is possible to point

[17] http://www-iepm.slac.stanford.edu.   [18] http://www.ripe.net.

out the presence of congestion points and bottlenecks. This information helps to identify specific areas in which routing policies and hardware modifications can lead to an improvement of the local traffic. At the global level, it is also possible to study the statistical fluctuations in performance at different points of the Internet and their correlation with geographic areas.

A different measure of network performance is obtained by looking at routing instabilities. A routing instability (also referred to as a *route flap*) is defined as a rapid change of network reachability. The origin of routing instabilities can be traced back to router congestions, transient physical outages, or configuration errors. When a router is unreachable, its peers send a BGP update, announcing the withdrawal of the corresponding routing paths. The message is then spread from peer to peer through a large number of BGP updates. The subsequent change in routing paths may create other flaps and the routing instability can propagate through the network and eventually lead to the transient loss of connectivity of large regions of the Internet. Noticeably, the occurrence and propagation of routing instabilities can be passively measured by collecting BGP updates. For instance, a large amount of data has been collected by the Internet Performance Measurement and Analysis (IPMA) project,[19] a joint effort by the University of Michigan Department of Electrical Engineering and the Computer Science and Merit Network. These data provide the basis for the study of the propagation of routing instabilities, and the impact of policy and topology on Internet routing performance (Labovitz, Malan, and Jahanian, 1998; Labovitz, Ahuja, Bose, and Jahanian, 2001).

We want to emphasize at this point that performance measurements are fundamental in providing a picture of the global Internet that goes beyond the connectivity layout obtained by studying the IR and AS level connectivity maps. The Internet's evolution and the various dynamical processes occurring in it are naturally constrained by the need for global efficiency in terms of traffic and performance. These elements are not usually considered in Internet modeling, that mainly focuses on the bare skeleton of the Internet graph. Indeed, the measurements of load, data traffic, and bandwidth at a global level are still at an infant stage because of the technical problems inherent in the traffic evaluation tools, which often impose a heavy load on the network, severely restricting their use. It is natural to expect that in the next few years this will become a large area of activity that will trigger a new understanding of the Internet's structure.

[19] http://www.merit.edu/ipma/.

4
The Internet's large-scale topology

We have seen in the previous chapter that the graphs representing the physical lay-out of the large-scale Internet look like a haphazard set of points and lines, with the result that they are of little help in finding any quantitative characterization or hidden pattern underlying the network fabric. The intricate appearance of these graphs, however, corresponds to the large-scale heterogeneity of the Internet and prompts us to the use of a statistical analysis as the proper tool for a useful mathematical characterization of this system. Indeed, in large heterogeneous systems, large-scale regularities cannot be found by looking at local elements or properties.[1] Similarly, the study of a single router connectivity or history will not allow us to understand the behavior of the Internet as a whole. In other words, we must abandon local descriptions in favor of a large-scale statistical characterization, taking into account the aggregate properties of the many interacting units that compose the Internet.

The statistical description of Internet maps finds its natural framework in graph theory and the basic topological measures customarily used in this field.[2] Here we shall focus on some metrics such as the shortest path length, the clustering coefficient, and the degree distribution, which provide a basic and robust characterization of Internet maps. The statistical features of these metrics provide evidence of the small-world and scale-free properties of the Internet. These two properties are prominent concepts in the characterization of complex networks, expressing in concise mathematical terms the hidden regularities of the Internet's structure. We shall address as well – where data allow – the dynamical evolution of the various measurements and their stability over time. Finally, we shall consider properties related to the existence of hierarchy, correlations, and geography in the Internet.

---

[1] The classical example of this fact in the field of physics is the theory of critical phenomena. From the simple Hamiltonian describing most models, it is difficult to imagine the rich behavior they show close to a critical point (Stanley, 1971; Binney, Dowrick, Fisher, and Newman, 1992; Yeomans, 1992).

[2] A detailed presentation of the main concepts of graph theory can be found in Appendix A1.

The data analysis presented here constitutes the starting point for any scientific approach to the Internet's evolution and structure. This will be particularly evident in the following chapter, where the statistical physics modeling of the Internet will necessarily lever on the empirical evidence and statistical laws provided here. A final warning to the reader concerns the datasets used in the analysis presented. We have collected and discussed the most recent Internet maps available in the literature. Data gathering projects, however, are continuously making public larger maps, and very likely more complete samples will be available by the time this book is published.

4.1 The growth of the Internet

We start by reviewing the properties of Internet graphs, analyzing the most basic set of standard metrics. In the following, unless otherwise stated, the IR graph corresponds to the data collected during October/November 1999 by the SCAN project with the *Mercator* software (Govindan and Tangmunarunkit, 2000). The AS level graph collected by the *Oregon route-views* and the extended AS+ graph collected by the Topology Project of the Computer Science Department at Michigan University (Qian *et al.*, 2002) are both dated May 26, 2001. The latter map is obtained by enlarging the *Oregon route-views* data set by external BGP information (see Chapter 3), and it will be used in the present chapter to assess the stability of the topological and statistical properties of Internet graphs upon increasing the completeness of the mapping process. First of all, we need to quantify the size of the graphs by counting the total number of vertices (nodes) N and edges (links) E, which are reported in Table 4.1 for all the different graphs considered.[3]

Then we can focus on local metrics by looking at the following properties of each vertex i (see Figure 4.1):

- The *degree* k_i defines the number of edges incident to the vertex i, i.e. the number of connections of that vertex with other vertices in the network.

Map	N	E	$\langle k \rangle$	$\langle c \rangle$	$\langle \ell \rangle$	$\langle b \rangle / N$
IR	228,263	320,149	2.8	0.03	9.5	5.3
AS	11,174	23,409	4.2	0.30	3.6	2.3
AS+	11,461	32,730	5.7	0.35	3.6	2.3

Table 4.1 *Average metrics of the IR, AS, and AS+ level graphs (See the text for the metrics' definitions)*

---

[3] In the case of the IR level graph it has been considered the giant component of the graph (see Appendix A1).

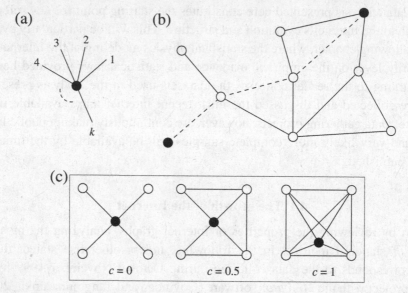

Fig. 4.1 Basic metrics characterizing a vertex i in the network. (a) The degree k quantifies the vertex connectivity. (b) The shortest path length identifies the minimum connected path (dashed line) between two different vertices. (c) The clustering coefficient provides a measure of the interconnectivity in the vertex's neighborhood. As an example, the central vertex in the figure has a clustering coefficient $c = 1$ if all its neighbors are connected and $c = 0$ if no interconnections are present.

- The ***shortest path length*** ℓ_{ij} between a pair of vertices i and j measures the number of edges forming the shortest path going from i to j.
- The ***betweenness*** b_i of a vertex i defines the total number of shortest paths among pairs of vertices in the network that pass through that vertex. If there are multiple shortest paths between a pair of vertices, the path passing through vertex i contributes to the betweenness with the corresponding relative weight (see Appendix A1).
- The ***clustering coefficient*** c_i of the vertex i is defined as the ratio between the number of edges e_i among its nearest neighbors and its maximum possible value $k_i(k_i - 1)/2$, i.e. $c_i = 2e_i/k_i(k_i - 1)$.

The degree and shortest path length of the vertices in IR (AS) level graphs have an immediate physical interpretation, quantifying how well a router (AS) is connected, and its minimum distance to other routers (ASs), in terms of router (AS) hops. The betweenness of a vertex gives a measure of the amount of traffic that goes through it, if the shortest path length is used as the metric defining the optimal path between pairs of vertices.[4] The betweenness, sometimes referred to as "load" (Goh, Kahng, and Kim, 2001), is thus a measure of the *centrality* of a vertex

---

[4] This is not always the case, since policy agreements can indeed impose non-optimal paths in the routing tables.

in the network. Furthermore, the clustering coefficient is a measure of the level of local interconnectivity among neighboring routers (ASs). For instance, a high clustering coefficient indicates a well interconnected local community of routers (ASs), very likely within the same administrative domain or geographical region.

All the defined metrics correspond to local properties of the vertices. In order to have a large-scale view of Internet graphs one has to shift the attention to statistical measures that take into account the global behavior of these quantities. A first global characterization of Internet maps can thus be obtained by measuring the statistical averages of each metric x_i over all the vertices in the network, $\langle x \rangle = N^{-1} \sum_i x_i$. In Table 4.1 we report the average values measured at both the IR and AS granularity.[5] The average values reported in Table 4.1 readily give some indications of the Internet's overall structure. The average degree of the three maps is very small if compared with the network sizes; they are therefore very *sparse* graphs. Despite this small average degree, however, the average shortest path length is also very small. This feature points towards the so-called *small-world* property, that will be discussed in detail in the next section. The differences among the metrics at different granularities are consistent with the fact that the AS and AS+ maps are coarse-grained representations of the IR map. The IR level map is sparser, and its average shortest path length is longer. Moreover, the IR map has a smaller average degree because routers in general support a limited number of connections (interfaces). On the contrary, ASs can have in principle a larger number of connections, since they represent the aggregation of several routers. Finally, both graphs have an appreciable number of vertices with a high number of connections (hubs), providing the shortcuts needed to generate a very small average shortest path length .

The National Laboratory for Applied Network Research (NLANR) (McGregor *et al.*, 2000) has been archiving since 1997 connectivity maps at the AS level, obtained by the *Oregon route-views* project. This effort is an invaluable service to the research community, allowing the analysis of the dynamical evolution of the AS level metrics. From the NLANR data we can observe that the dynamical evolution of Internet growth is a non-trivial process, which is not simply driven by the addition of new ASs. Inspecting the daily database, it is possible to check that changes in the maps are due to both the addition (birth) and deletion (death) of ASs and their peering relations (Pastor-Satorras, Vázquez, and Vespignani, 2001; Qian *et al.*, 2002; Vázquez, Pastor-Satorras and Vespignani, 2002b). In Figure 4.2 we show the monthly number of new and deleted vertices in the range between November 1998 and November 2000 (Qian *et al.*, 2002). This plot has been constructed analyzing

---

[5] Note that, for the case of the clustering coefficient, it is possible to define a *restricted* average over the vertices with degree $k_i \geq 2$ (Jin and Bestavros, 2002; Bu and Towsley, 2002). This restricted average yields values larger that those reported in Table 4.1.

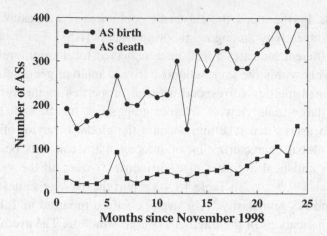

Fig. 4.2   Monthly number of new and dead ASs in the period November 1998 to November 2000. Data from Qian *et al.* (2002).

limited maps from the data obtained from the eight routers that have maintained a steady relationship with the Oregon route server during the period considered. Despite their smaller sizes, these limited maps have the advantage of considering always the same view of the Internet, and are thus less prone to false birth/death events due to temporal changes in the set of peer routers. This and other parallel analyses (Pastor-Satorras *et al.*, 2001; Vázquez *et al.*, 2002b) provide evidence that the AS's birth rate is much larger than the corresponding deletion rate. More interestingly, if we restrict our attention to the fraction of vertices deleted or added as a function of their degree, Table 4.2, we observe that only poorly connected nodes have an appreciable probability to disappear, while it is very unlikely to witness the addition of a heavily connected new AS. This fact is easily understandable in terms of the market competition among ISPs, where the largest fraction of new vertices correspond to small, little connected newcomers, which are at the same time the ones that more likely go out of business due to the market pressure.

Similarly, it is possible to keep track of the number E_{new} of edges appearing between a newly introduced vertex and vertices already present, and the number E_{old} of edges between already existing vertices (Vázquez *et al.*, 2002b). From Table 4.3 we can observe that the rate of creation of new edges is governed by these two processes at the same time. Specifically, the growth of the Internet's connectivity is dominated by the appearance of edges between already existing vertices. This fact indicates that Internet growth is strongly driven by the need for wiring redundancy and increasing available bandwidth for data transmission. A more detailed analysis (Qian *et al.*, 2002) also allows us to distinguish between *customer* and *peer* wiring events, in which a new edge is established between a small degree vertex (the customer AS) and a large degree vertex (the provider AS) or between vertices

Degree	Number of dead ASs	Number of new ASs
1	1,184	5,591
2	204	816
3	22	23
4	6	4
5	4	1
6	1	1
7	1	1
9	1	—
10	—	1
11	—	1
12	1	—
14	—	1
48	1	—

Table 4.2 *Total number of new and dead ASs in the period November 1998 to November 2000 (Data from Qian et al., 2002)*

Year	1998	1999
E_{new}	170(8)	231(11)
E_{old}	350(9)	450(29)
E_{new}/E_{old}	0.48(2)	0.53(3)

Table 4.3 *Monthly rate of new edges connecting existing vertices to new (E_{new}) and old (E_{old}) vertices, in the years 1998 and 1999 (Data from Vázquez et al., 2002)*

with similar degrees, respectively. The identification of the so-called *rewiring processes* is also very interesting. Rewiring an edge refers to an event in which a specific AS shifts one of its previous connections to another existing AS. Empirically, a birth/death process amounts to a rewiring process if it happens within a rather short time scale. Qian *et al.* (2002) consider that a rewiring process occurs if an edge birth happens within ten days of an edge disappearance (after or before) on the same vertex. Still, birth/death processes might be just random coincidences and it is possible to show that only 20% of all birth/death processes occurring on the same vertex are meaningfully correlated, particularly if they occur within a one day time window.

N	3,700	4,500	5,200	6,300	6,500	8,900	9,100	10,500
$\langle k \rangle$	3.6	3.7	3.8	3.9	3.9	4.0	4.1	4.1
$\langle c \rangle$	0.21	0.23	0.24	0.25	0.25	0.29	0.29	0.29
$\langle \ell \rangle$	3.7	3.7	3.7	3.7	3.7	3.7	3.7	3.7
$\langle b \rangle / N$	2.4	2.4	2.4	2.4	2.4	2.3	2.3	2.3

Table 4.4 *Behavior of the average basic metrics as a function of the number N of ASs recorded in the NLANR maps*

The empirical observations reported so far show that the overall growth of the Internet is the outcome of the net balance of a birth/death dynamics involving large fractions of the system. However, despite these complex dynamics and the overall growth of vertices and edges, the behavior in time of the various metrics characterizing Internet graphs shows much smaller variations. Indeed, in Table 4.4 we observe that, in the time span considered, the average values of all the metrics do not show large fluctuations and seem to approach a stationary value at which statistical properties are fairly stable in time. This evidence allows us to use the statistical properties previously defined in the characterization of the Internet. They do not suffer from relevant changes with network size, and can be considered as inherent properties of the Internet. Over the following sections we shall leverage on these statistical properties to express in mathematical terms the complex patterns emerging from the Internet maps.

4.2 Small-world properties

4.2.1 Degrees of separation

The average shortest path length among vertices found in Internet maps is very small if compared with the size of the graphs. This empirical evidence was reported in early analysis of Internet data (Faloutsos, Faloutsos and Faloutsos, 1999), and it has been confirmed for all recent datasets (Huffaker *et al.*, 2000; Huffaker, Fomenkov, Moore, Plummer and Claffy, 2002; Pastor-Satorras *et al.*, 2001; Qian *et al.*, 2002; Bu and Towsley, 2002). In particular, several measurements focused on the IP path hop count; i.e. the number of IP addresses traversed by an IP packet along the path between a source-destination host pair. While the actual IP hop count is not always equivalent to the shortest path length due to routing policies, small hop counts are consistently found in hosts all around the world.[6]

---

[6] The CAIDA group reported in the first five months of 2001 an average hop distance $L_{IP} = 15.3 \pm 4.2$ between their monitors and 313, 471 destinations in the IP space (Huffaker *et al.*, 2002)

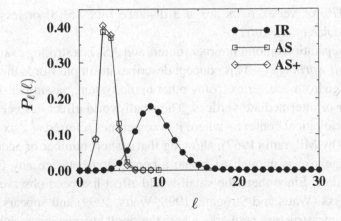

Fig. 4.3 Probability distribution $P_\ell(\ell)$ of the minimum path distance between vertices, for the AS, AS+, and IR graphs. The distance is measured as the number of edges traversed from the source vertex to the destination vertex.

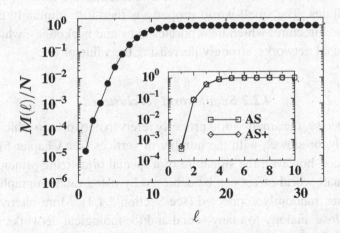

Fig. 4.4 Hop plots $M(\ell)$ corresponding to the AS, AS+, and IR graphs.

A more precise characterization of the mild variation in the path distance is provided by the analysis of the probability $P_\ell(\ell)$ of finding two vertices separated by a shortest path length ℓ, Figure 4.3. For the IR, AS, and AS+ graphs the distribution is sharply peaked around the average value $\langle \ell \rangle$, which can be considered as the *characteristic* length of the networks. The distribution drops very rapidly away from the peak value, as can be better seen from the *hop plot* $M(\ell)$, Figure 4.4, first introduced by Faloutsos *et al.* (1999), and defined as the average number of vertices within a distance less than or equal to ℓ for any given vertex (see Appendix A1). From Figure 4.4, for instance, it is possible to check that at the IR

(AS) level, 97% of vertex pairs are at a distance ℓ of 15 (8) or less (Vázquez *et al.*, 2002b; Qian *et al.*, 2002).

The small separation among Internet routers and ASs is a striking example of the so-called *small-world effect*. This concept describes in simple words the fact that it is possible to go from one vertex to any other in the system, passing through a very small number of intermediate vertices. The small-world effect has been popularized in the sociological context – where it is sometimes referred as "six degrees of separation" – by Milgram (1967), showing that a short number of acquaintances (on the average six) is enough to create a connection between any two people chosen at random. Since then, the small-world effect has been observed in many natural networks (Watts and Strogatz, 1998; Watts, 1999) and appears to characterize several infrastructure networks where the small average distance is crucially important to speed up communications (Bush, Files, and Thompson, 2001). For instance, if the Internet had the shape of a regular grid, its characteristic distance would scale as $\langle \ell \rangle \sim N^{1/2}$; with the present Internet size, each IP packet would pass through 10^3 or more routers, drastically depleting the Internet's communication capabilities. The small-world property is therefore implicitly enforced in the network architecture, which incorporates hubs and backbones, which connect different regional networks, strongly decreasing the value of $\langle \ell \rangle$.

4.2.2 Small-world yet clustered

To be more precise, the small-world property refers to networks in which $\langle \ell \rangle$ scales logarithmically, or slower, with the number of vertices (see Chapter 5). This feature alone is not, however, the signature of a special organizing principle. For instance, the small-world effect can be achieved by using random graphs in which vertices are just randomly connected (see Section 5.1.1). More interesting is the fact that, in close analogy to many social and technological networks (Watts and Strogatz, 1998; Watts, 1999), the small-world effect goes along with a high level of clustering. The clustering coefficient quantifies the fraction of pairs of neighbors of a vertex which on their turn are neighbors of each other. In a random graph with N vertices and average degree $\langle k \rangle$, the average clustering coefficient is $\langle c \rangle \sim N^{-1}$ (see Section 5.1.1). The clustering coefficients for the AS, AS+, and IR maps listed in Table 4.1 are two to four orders of magnitude larger than the corresponding value for a random graph of the same size (vertices and edges). This large clustering coefficient corresponds in the Internet to a large statistical abundance of "communities" in which every router (AS) is connected to every other. This fact is in turn related to both hierarchical and geographical factors that will be addressed in the following sections. On the other hand, the finding of clustered networks with small-world properties raises a very interesting issue: Random graphs feature the small-world effect but are not clustered, while regular grids tend to be clustered but are not small-world. Is there a different kind of network, driven by different organizing principles, able to capture both properties at the same time?

4.3 Heavy tailed distributions

4.3.1 Degree fluctuations and power-law behavior

The analysis of the average metrics presented in the previous sections rules out the possibility of a purely random graph structure or a regular grid architecture. In addition, even a visual inspection of the maps clearly shows a very high level of heterogeneity in the connectivity properties (see for instance Figure 3.7). Many vertices in the maps have just a few connections, while a few hubs collect hundreds or even thousands of edges. Somehow, we cannot infer large-scale properties from the very fluctuating local ones. The evidence suggests a peculiar topology that will be more clearly identified by looking at the detailed statistical distributions of the different metrics. In particular, Faloutsos *et al.* (1999) pointed out for the first time that the connectivity properties of the IR and AS level maps are characterized by *heavy tailed* probability distributions that can be reasonably approximated by power-law forms. For instance, Figure 4.5 reports the probability $P(k)$ that any given vertex in the IR level map has degree k. The distribution is highly variable in the sense that degrees vary over a range close to three orders of magnitude. This behavior is very different from the case of the bell-shaped, exponentially decaying distributions shown in Figure 4.3. At the AS level the same heavy tailed degree distribution is observed. Figure 4.6 shows the cumulative degree distribution $P^c(k)$

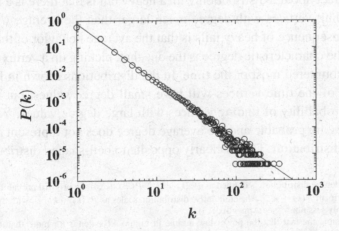

Fig. 4.5 Degree distribution $P(k)$ for the IR level graph in a double logarithmic scale. The solid line is a power law decay $k^{-\gamma}$ with $\gamma = 2.1$.

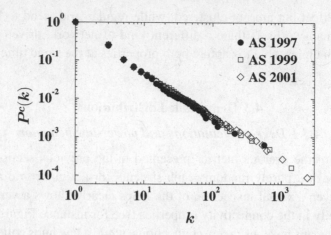

Fig. 4.6 Cumulative degree distribution for the *Oregon route-views* AS graphs in the years 1997, 1999, and 2001. The power-law behavior is characterized by a slope -1.1, which yields a degree exponent $\gamma \simeq 2.1$.

of the AS maps obtained from the *Oregon route-views* for three different years.[7] The AS level degree distribution is stable over the years and only the cut-off, fixed by the maximum degree of the system, is increasing due to overall Internet growth. At both the AS and IR level, the degree distribution appears to be well approximated by the linear behavior on the double logarithmic scale.[8] More precisely, the distribution can be fitted in a wide region of k values by the power-law form[9]

$$P(k) \simeq ak^{-\gamma}, \tag{4.1}$$

with exponent $\gamma \simeq 2.1$ and an opportune normalization constant a.

A peculiar fact about a distribution with a heavy tail is that there is a finite probability of finding vertices with degree much larger than the average $\langle k \rangle$. In other words, the consequence of heavy tails is that the average behavior of the system is not typical. The characteristic degree is the one that, picking up a vertex at random, should be encountered most of the time. In the distributions shown in Figures 4.5 and 4.6 most of the time vertices will have small degree values, but there is an appreciable probability of finding vertices with large degree values. Yet all intermediate values are probable and the average degree does not represent any special value for the distribution. This is clearly opposite to bell-shaped distributions with

---

[7] The cumulative degree distribution is defined as $P^c(k) = \int_k^\infty P(k')\, dk'$. If the original probability distribution has a power-law form $P(k) \sim k^{-\gamma}$, the cumulative distribution scales as $P^c(k) \sim k^{1-\gamma}$, with the advantage of being considerably less noisy (see Appendix A3).

[8] It is most noticeable the fact that the power-law regime in Figure 4.6 extends for more than three orders of magnitude. Such a large range is not often encountered in real physical systems.

[9] Power-law distributions are sometimes referred to as Pareto distributions, after the name of the economist that observed this behavior in the income distribution (Pareto, 1896).

fast decaying tails, in which the average value is very close to the maximum of the distribution and represents the most probable value in the system. The power-law behavior and the relative exponent thus represent a quantitative measure of the level of heterogeneity of the network's connectivity.

In more mathematical terms the heavy-tail property translates into a very large level of degree fluctuations. This can be observed by inspecting the normalized variance of the distribution, $\sigma^2/\langle k \rangle^2$. In the case of distributions with a power-law tail with exponent $2 < \gamma < 3$ we have that

$$\langle k \rangle = \int_m^\infty k P(k) \, \mathrm{d}k = \text{const.,} \tag{4.2}$$

where $m \geq 1$ is the lowest possible degree in the network.[10] The average degree is therefore well defined and bounded. However, the variance $\sigma^2 = \langle k^2 \rangle - \langle k \rangle^2$ is dominated by the second moment of the distribution, that diverges with the upper integration limit k_c as

$$\langle k^2 \rangle \sim \int k^2 P(k) \, \mathrm{d}k \sim \int_m^{k_c} k^{2-\gamma} \, \mathrm{d}k \sim k_c^{3-\gamma}. \tag{4.3}$$

In the asymptotic limit $k_c \to \infty$, fluctuations are therefore unbounded and depend only on the system size.[11] The absence of any intrinsic scale for the fluctuations implies that the average value is not a characteristic scale for the system. In other words, we are in the presence of a *scale-free* network for what concerns the statistical properties of the vertices' degree. This reasoning can be extended to values of $\gamma \leq 2$, since in this case even the first moment is unbounded.

The absence of an intrinsic characteristic scale in a power-law distribution is also reflected in the self-similarity properties of such distribution; i.e. it looks the same at all length scales. This means that if we look at the distribution of degrees by using a coarser scale in which $k \to \lambda k$, with λ representing a magnification/reduction factor, the distribution would still have the same form. This is not the case if a well-defined characteristic length is present in the system.

Finally, from the previous discussion, it is possible to provide a heuristic characterization of the level of heterogeneity of networks by defining the parameter

$$\kappa = \frac{\langle k^2 \rangle}{\langle k \rangle}. \tag{4.4}$$

Indeed, fluctuations are denoted by the normalized variance that can be expressed as $\kappa/\langle k \rangle - 1$, and scale-free networks are characterized by $\kappa \to \infty$, whereas

---

[10] For the sake of simplicity we consider a continuous variable k (see Section 5.4). The same results hold in the discrete case, where the integral is replaced by a summation.

[11] In statistical physics the infinite size limit is called the *thermodynamic limit*.

homogeneous networks have $\kappa \sim \langle k \rangle$. For this reason, we will generally refer all networks with heterogeneity parameter $\kappa \gg \langle k \rangle$ as scale-free networks.[12] We shall see in the following chapters that κ is a key parameter for all properties and physical processes in networks which are affected by the degree fluctuations.

4.3.2 Further evidence of scale-free behavior

The possibility that the Internet belongs to the class of scale-free networks, and the eventual consequences of the scale-free behavior, has impacted on the networking community and triggered off, after the first observations by Faloutsos *et al.* (1999), a significant amount of activity aimed at the confirmation of this finding. Further empirical evidence for the heavy-tailed behavior of the degree distribution has since been collected both at the IR and AS level in several studies (Govindan and Tangmunarunkit, 2000; Tangmunarunkit, Doyle, Govindan, Jamin, Shenker and Willinger, 2001; Broido and Claffy, 2001; Pastor-Satorras *et al.*, 2001; Magoni and Pansiot, 2001; Chang, Jamin, and Willinger, 2001; Vázquez *et al.*, 2002b; Tangmunarunkit, Govindan, Jamin, Shenker and Willinger, 2002*a*; Willinger, Govindan, Jamin, Paxson and Shenker, 2002; Qian *et al.*, 2002; Yook *et al.*, 2002; Bu and Towsley, 2002). It must be noted that many of these works report as evidence of the scale-free behavior of the degree distribution the power-law form of Zipf's plot. This is obtained by assigning a rank r to the vertices in terms of their degree value. The vertex with the largest degree has rank $r = 1$, and so on in decreasing degree order. Zipf's ranked distributions plot the degree k of vertices with respect to their rank r. The plots obtained have a power-law behavior $k \sim r^{-\alpha}$ that is common to many physical phenomena (Zipf, 1949; Faloutsos *et al.*, 1999). This power-law behavior is indeed a direct consequence of the power-law behavior of the degree distribution. The probability of having a vertex with a degree larger than k is given by the cumulative distribution $P^c(k) \sim k^{-\gamma+1}$. In a graph with N vertices the rank of a vertex of degree k is therefore $r \sim Nk^{-\gamma+1}$. Inverting this last relation yields the Zipf's ranked distribution with $\alpha = (\gamma - 1)^{-1}$. This shows that the two power-law exponents are not independent but provide different representation of the same scale-free feature.

The scale-free behavior, in fact, does not show up only in the degree distribution. For instance, additional evidence can be observed in the power-law behavior of degree correlation functions, as we shall see in Section 4.5. Furthermore, the betweenness distribution $P_b(b)$ (i.e. the probability that any given vertex is traversed by b shortest paths between pairs of vertices) shows heavy tails with scale-free behavior. This latter fact is made evident in Figure 4.7, where we plot, for the IR, AS,

---

[12] Obviously, in the real world κ cannot be infinite since it is limited by the network size.

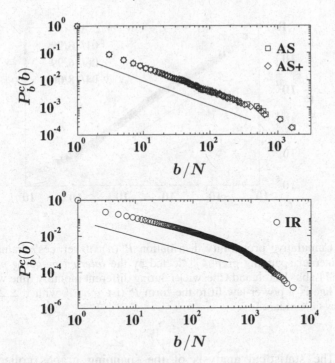

Fig. 4.7 Cumulative betweenness distribution $P_b^c(b)$ for the AS, AS+, and IR graphs. The solid line is a power law decay $P_b^c(b) \sim b^{1-\gamma_b}$ with $\gamma_b \simeq 2.0$.

and AS+ maps, the cumulative betweenness distribution $P_b^c(b) = \int_b^\infty P_b(b')\,db'$. The betweenness distribution of the AS and AS+ maps is almost identical and well approximated by the power law behavior $P_b(b) \sim b^{-\gamma_b}$ with $\gamma_b \simeq 2.0$. At the IR granularity, the betweenness distribution varies over three orders of magnitude with an initial power-law behavior followed by a faster decay at large values of the betweenness. The large value truncation finds its origin in the routers' finite capacity, which is discussed in detail in the next section.

A different manifestation of the scale-free nature of the Internet is provided by the analysis of spanning tree graphs obtained by single probe experiments (Caldarelli, Marchetti, and Pietronero, 2000). In this case, the probe host is considered as the outlet of a river basin and the path connecting this point to all the possible detected IP addresses can be considered to form the structure of this basin (Rodriguez-Iturbe and Rinaldo, 1997). An interesting measure, customarily used in river network studies, is the probability density function $P(n)$ that any vertex in the basin connects n other vertices uphill. The quantity n is usually referred to as the *drainage area*, and in the path-probing language corresponds to the number of IP addresses reachable from the probing host by paths traversing

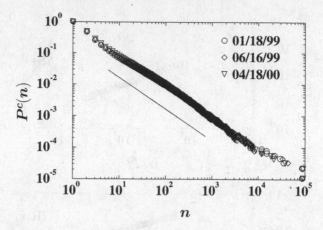

Fig. 4.8 Cumulative probability distribution $P^c(n)$ of vertices' drainage area n in the Internet spanning graphs collected by the *Internet mapping project* at Lucent Bell Labs (Burch and Cheswick) during different monthly time windows. The solid line is a power-law fit to the form $P^c(n) \sim n^{1-\tau}$ with $\tau \simeq 2.0$. Data provided by G. Caldarelli.

that vertex. The statistical analysis of the spanning graphs collected by the *Internet mapping project* at Lucent Bell Labs (Burch and Cheswick, 1999) reveals a clear power law distribution $P(n) \sim n^{-\tau}$ with $\tau \simeq 2.0$ (see Figure 4.8). This exponent is different from those found in river networks ($\tau \simeq 1.45$) and uncorrelated branching processes ($\tau = 1.5$) (Rodriguez-Iturbe and Rinaldo, 1997), and not surprisingly, it is in good agreement with the betweenness exponent γ_b. Somehow, the drainage area is just the betweenness of a vertex restricted only to the shortest paths connecting all the vertices with the probing host. The equivalence of the two exponents should be expected in a system that has self-averaging properties; i.e. the statistical properties of the system are independent of the specific observation point. This is not an obvious fact, and should not be confused with a homogeneity assumption. On the contrary, it states that whatever reference vertex we use for measuring the statistical properties, we find the same amount of fluctuations.

Table 4.5 summarizes the numerical properties of the various heavy tailed probability distributions analyzed so far. The scale-free behavior is especially evident when comparing the heterogeneity parameter κ and the wide variations in the variables range. For the sake of comparison we also report the same properties in the case of the short-tailed distribution, observed for the shortest path length in the various samples.

Finally, it is important to mention that other interesting power-law relations are found by looking at the *spectral analysis* of Internet graphs. This analysis yields

Variable x	Sample	$\langle x \rangle$	x_{max}	σ	κ	exponent
	IR	2.8	1,937	8.4	28	2.1 ± 0.1
k	AS	4.2	2,389	33.0	264	2.1 ± 0.1
	AS+	5.7	2,432	35.1	222	2.2 ± 0.1
	IR	5.3	6,878	57.5	249	2.0 ± 0.2
b/N	AS	2.3	1,819	22.9	229	2.0 ± 0.1
	AS+	2.3	1,795	21.9	213	2.0 ± 0.1
n	Lucent	10.9	91,419	467	19,991	2.0 ± 0.2
	IR	9.5	32	2.37	10.1	∞
ℓ	AS	3.6	10	1.45	4.2	∞
	AS+	3.6	10	1.51	4.2	∞

Table 4.5 *Numerical summary of the probability distributions analyzed in this chapter. x_{max} is the maximum value of the variable observed in the sample. The parameter $\kappa = \langle x^2 \rangle / \langle x \rangle$ and the mean square root deviation σ estimate the level of fluctuations in the sample. It is possible to appreciate that all heavy-tailed distributions show a maximum value of the variable $x_{max} \gg \langle x \rangle$ and a heterogeneity parameter $\kappa \gg \langle x \rangle$ (at least one order of magnitude larger). This is not the case for the shortest path length distributions, where $\kappa \simeq \langle x \rangle$. For the sake of comparison we reported the values $\langle x \rangle$ already presented in Table 4.1. The exponent is evaluated by a best fit procedure in the linear region of the double logarithmic plot. The value ∞ indicates a decay faster than any power-law*

a structural classification of the Internet's networks defining subgraphs of different sizes and connectivity properties depending on the multiplicity of the eigenvalues of adjacency-like matrices (see Appendix A1). Noticeably, the statistical analysis of the occurrence of subgraphs and their interconnections reveals several interesting power-law relations (Faloutsos *et al.*, 1999; Vukadinovic, Huang, and Erlebach, 2002).

4.4 Critically examining scale-free properties

Along with the more extensive maps available and the more detailed analyses performed, the scale-free behavior of the Internet has been carefully scrutinized and some questions have been raised in the literature (Broido and Claffy, 2001; Qian *et al.*, 2002; Vázquez, Pastor-Satorras, and Vespignani, 2002a). It is therefore important to discuss the various objections to scale-free behavior that can be found in the literature.

4.4.1 Size and physical constraints

A first relevant criticism to the observed scale-free behavior of the Internet concerns the fact that, in some cases, deviations and smoothing out of the pure power-law behavior are found in the analysis of statistical distributions. For instance, in Figure 4.5 the power-law fit of the form $P(k) = ak^{-\gamma}$ with exponent $\gamma = 2.1 \pm 0.1$ (a is a normalization constant) is fairly consistent for $k < 60$. For $k > 60$, however, one can observe that the degree distribution follows a faster decay. By using the cumulative distribution or binning procedures to reduce noise (see Appendix A3), this effect becomes more evident and the power-law is smoothed by a cut-off regime. This picture is consistent with a finite size scaling of the form $P(k) = k^{-\gamma} f(k/k_c)$ (Dorogovtsev and Mendes, 2002), where $f(x)$ has the asymptotic behavior $f(x) = $ const. for $x \ll 1$ and $f(x) \ll 1$ for $x \gg 1$. Here k_c is the degree above which the distribution decays faster than a power law, and represents the natural integration limit in Equation (4.3). Deviations from the power-law behavior at large connectivities are also clearly observed for the larger IP level maps analyzed by Broido and Claffy (2001).[13]

The presence of truncations in power laws, however, should not be considered a surprise, since it finds a natural place in the context of scale-free phenomena. Actually, the heavy tail truncation is the natural effect of the upper limit of the distribution that must necessarily be present in every real-world system. Indeed, bounded power laws (i.e. power-law distributions with a cut-off) are also observed in other real networks (Amaral, Scala, Barthélémy, and Stanley, 2000) and different mechanisms have been proposed to account for the presence of large degree truncations. Specifically, we can distinguish two different kinds of cut-offs in real networks. The first is an exponential cut-off, $f(x) = \exp(-x)$, which can be explained in terms of a finite connectivity capacity of the network elements (Amaral *et al.*, 2000) or incomplete information (Mossa, Barthélémy, Stanley, and Amaral, 2002). In this case the cut-off value k_c is a constant that depends upon the external physical and technological constraints. It is likely that is what is happening at the IR level, where it is unlikely and technically unpractical to have an excessively large number of interfaces in a single router. This is a finite capacity constraint that, in our opinion, is the dominant mechanism affecting the tail of the IR degree distribution. From this perspective, larger and more recent samples at the IR level could present some shift in the cut-off, due to the improved technical router capabilities. A second possibility is given by a very steep cut-off such as $f(x) = \theta(1 - x)$, where $\theta(x)$ is the Heaviside step function.[14] This is what happens in growing

---

[13] In this work, however, the cumulative degree distribution was fitted to a Weibull distribution $P^c(k) = a \exp[-(k/k_c)^\beta]$.

[14] The Heaviside step function takes the value $\theta(x) = 0$ for $x < 0$, and $\theta(x) = 1$ for $x > 1$.

networks with a finite number of elements. Since scale-free networks are often dynamically growing networks, this case represents a network which has grown up to a finite number of vertices N. The maximum degree is therefore a function of the networks size $k_c(N) < N$ (see Appendix A5). The scale-free behavior of the degree distribution is evident up to $k_c(N)$, above which it decays as a step function, since the network does not possess any vertex with degree larger than $k_c(N)$. Inspecting Figure 4.6, this second possibility appears to be realized at the AS level. Interestingly, Tangmunarunkit *et al.* (2001) reported that the size of ASs in terms of routers is linearly correlated with their measured degree. This fact indicates that the dominant mechanism at the AS level is the finite size of the network, while degree limits are not present, since large ASs can handle a very large connectivity load.

The connection between finite capacity and bounded distributions also becomes evident if we consider the betweenness. This magnitude is a static estimate of the amount of traffic that a vertex supports. Hence, if a router has a bounded capacity, the betweenness distribution should also be bounded for large betweenness. However, this effect should be absent for the AS maps. The cumulative betweenness distribution $P_b^c(b)$ for the AS, AS+, and IR maps is shown in Figure 4.7. The AS and AS+ distributions are practically equal, with an exponent $\gamma_b \simeq 2.0$. In the case of the IR map, however, the betweenness distribution follows a truncated power law, in analogy to the behavior observed for the degree distribution.

4.4.2 Statistical reliability

A more radical objection is related to the possibility that the power-law behavior could be an artifact of the map's incompleteness. As we discussed in Chapter 3, the collection of Internet maps suffers from biases at all granularity levels, and it is not possible to ascertain precisely to what extent they provide a reliable picture of the Internet's connectivity. The number of vertices reported depends on the various collection methods, which in some cases is known to be very close to the full number of vertices available (Broido and Claffy, 2001; Qian *et al.*, 2002; Huffaker, Fomenkov, Moore, Plummer, and Claffy, 2002). Concerning the interconnectivity (edges), the situation is quite different, since the achieved level of sampling is not known *a priori*. It is then natural to wonder if this incomplete information might affect the statistical analysis, yielding spurious results. This question does not find an easy answer. The reliability of the graphs depends on the number of probing stations and targeted routers used to infer the physical connectivity (Barford, Bestavros, Byers, and Crovella, 2001) (see Chapter 3). In the case of a very incomplete sampling, it has been shown that statistical distributions can develop an apparent long tail behavior even if the graph topology is not scale-free (Lakhina,

Byers, Crovella, and Xie, 2002). It is possible to show, however, that the high clustering coefficient observed in Internet graphs is generally associated with a fair statistical sampling of the network interconnectivity, since a large enough number of closed triangles must be detected in the graph in order to produce a substantial value of $\langle c \rangle$ (Vázquez *et al.*, 2003). This evidence, along with the very large maximum degree of the graphs and the actual level of sampling achieved by most recent projects (almost 70% of the addressable networks in the Internet) supports the view that the highly variable distributions are a genuine feature of the Internet.

In this respect, Qian *et al.* (2002) obtained an important result by performing an empirical analysis of the completeness of the BGP derived maps reconstructed from the *Oregon route-views* and the reliability of the resulting graphs at the AS level. Their study reveals that, being strictly BGP-based, the data provide a rather incomplete picture of the Internet's connectivity. The *Oregon route-views* collects and merges the connectivity inferred from the BGP routing tables of 20 to 50 operational routers. Each of them provides its routing tables, but for several reasons a large amount of interconnections are missing.[15] In particular, these maps typically miss approximately more than 30% of the physical links that might be obtained by using additional sources in the construction of the AS graphs. Indeed, by using supplementary BGP information from a large number of Internet ASs and Internet Routing Registries, Qian *et al.* (2002) constructed the extended AS+ graph, which contains only 2% more vertices (ASs) but almost 40% more edges (peering connections), as reported in Table 4.1. Obviously, such a conspicuous difference is very relevant in a detailed quantitative analysis of the graph. It will provide very different views of local regions of the Internet and the associated connectivity. However, this result does not necessarily imply a different qualitative behavior of the large-scale statistical properties.

In Figure 4.9 we compare the connectivity distribution for the AS and AS+ maps collected on the same date. Strikingly, the degree distribution obtained for the AS+ map is clearly highly variable with a power-law tail very similar to that observed in the more incomplete AS map. This is an extremely important result since the degree distribution could be strongly affected by such a large increase in the number of edges, especially if the sampling of standard maps suffered from unpredictable biases due to routing policies or specific locations of the collecting routers. Not surprisingly, a very careful inspection shows deviations from a pure power law at intermediate degrees in the AS+ map, and this anomaly might or might not be related to the biased enlargement of the Internet sampling (Qian *et al.*, 2002). However, while this represents an important point in the detailed description of the connectivity properties, it is not crucial in what concerns the large-scale nature of

---

[15] See Chapter 3 for a detailed discussion of mapping limitations.

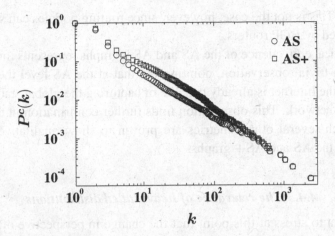

Fig. 4.9   Cumulative degree distribution $P^c(k)$ for the AS *Oregon route-views* graph and the extended AS+ graph, both collected on May 26, 2001.

the Internet. This fact emerges clearly by looking at Table 4.5, where it is possible to appreciate the similar features of both statistical distributions. For instance, with respect to the network's physical properties, it is just the large connectivity region that is actually effective. Indeed, the heterogeneity parameter κ, that controls the network robustness to removal of vertices (see Chapter 6) and spreading phenomena (see Chapter 9), is mainly determined by the tail of the distribution, and is very similar for both maps. In particular, we estimate $\kappa = 264$ and $\kappa = 222$ for the AS and AS+ maps, respectively. With such large values of κ, for all practical purposes (resilience, virus spreading, traffic, etc.) the AS and AS+ maps behave qualitatively in very similar ways.

Equally striking is the evidence purported by the scale-free properties of the betweenness distribution, which is very similar for both AS and AS+ graphs as shown in Figure 4.7. In this case no differences can be detected by visual inspection, and a clear power-law behavior is observed for both distributions. The equivalence of the AS and AS+ graphs in respect to the betweenness distribution is even more striking when we consider the origin of these graphs. The AS graph is constructed on the exclusive base of BGP routing tables. The routing paths traced in these tables represent a compromise between the shortest path length among different routers and the policy agreements reached by the different ISPs. On the other hand, the additional edges introduced in the AS+ graph correspond to alternative connections which are not frequently announced by the BGP peers probed by the *Oregon route-views* server. If the AS map connectivity were exclusively established on a shortest path length basis, it would not be surprising at all to observe that the addition of new connections does not change the betweenness distribution. This is not the case, however, since routing policies can strongly bias the paths layed by BGP routers.

The statistical equivalence of the AS and AS+ graphs represents thus a remarkable and non-trivial observation, pointing out that at the AS level the large-scale sampling of the Internet is already capable of capturing the global statistical properties of this network. This observation finds further confirmation in the next section, in which several others metrics are proven to show qualitatively identical properties in the AS and AS+ graphs.

4.4.3 *The relevance of heavy tailed distributions*

It is important to stress at this point that the change in perspective offered by the presence of scale-free distributions has an impact that goes far beyond the discussion concerning truncations and finite size effects, deviations from pure power-laws, or the exact values of the exponents. Indeed, fittings to a power-law form can yield slightly different results, depending on the range of values actually considered for the fit. In this sense, the exact value of the exponents is of secondary importance, at least given the amount of noise in the present datasets. The really crucial issue is that the observation of heavy tailed, highly variable distributions provides new experimental input for a radical change of the Internet representation and modeling.

Internet topology generators aim at representing the Internet by constructing algorithms that incorporate the desired topological properties of the graph (see Chapter 5). For years, topology generators have been inspired by homogeneous random graph models (see Section 5.1), yielding synthetic networks with Poisson degree distributions. However, the ubiquitous presence of heavy tailed distributions makes these generators inadequate. The differences introduced by power-laws and large fluctuations in many properties of the graph are too significant to be ignored: protocols designed for homogeneous random graphs perform badly on networks with heterogeneous connectivity patterns; resilience to damage is extremely different in networks with heavy tailed degree distribution; spreading and searching processes are greatly affected by degree fluctuations. These considerations motivate the development of new Internet topology generators that incorporate connectivity patterns in agreement with the empirical observations.

Looking at the scale-free properties from a wider aspect makes the Internet one of the most prominent examples of what appears as a general class of networks. Scale-free degree distributions and heterogeneity have indeed been observed for a large class of real-world networks belonging to the biological and social realms as well as to the technological domain. In Chapter 7 we shall see that many virtual and social networks (World Wide Web, peer-to-peer, e-mail) living in the Internet have scale-free properties. Similar features are also observed in the science collaboration graph (Newman, 2001a; Newman, 2001b) and the scientific citation network (Redner, 1998). Metabolic (Jeong, Tombor, Albert, Oltvai, and Barabási, 2000) and protein interaction networks (Jeong, Mason, Barabási, and Oltvai, 2001) are examples of scale-free networks from the biological world. The evidence that a scale-free topology is shared by many complex evolutive networks cannot be considered as incidental. Rather, it points to the possibility of some general principle that can possibly explain the emergency of this architecture in such different contexts (Dorogovtsev and Mendes, 2003).

With this picture in mind, it is possible to think on the more ambitious goal of modeling complex networks, and thus the Internet, by understanding the main dynamical principles at the basis of their evolution. At this level the presence of heavy tailed distributions represents the fundamental signature of an emergent cooperative behavior. In other words, the study of the collective behavior of the many elements forming the network can shed light on the large-scale structures in which it is eventually self-organized. As we shall see in the next chapter, this evidence calls for a different framework in network modeling.

4.5 The hierarchical structure of the Internet

The presence of a scale-free degree distribution does not necessarily imply a precise structural organization or hierarchy. At the same time, we know that the Internet is organized around some primary structure, such as the distinction between *stub* and *transit* domains.[16] This hierarchical organization is at the origin of the high clustering coefficient of the Internet, but its relation with the power-law behavior of the degree distribution is not obvious. In addition, while the IR and AS graphs represent the same overall structure of the Internet, the IR level contains a much lower degree of structural and administrative organization than the AS level, which is the outcome of a coarse graining strongly based on a modular construction. It is not clear, therefore, how much the two levels could have in common in a hierarchical analysis.

4.5.1 Up–down organization

A heuristic way to quantify the level of hierarchy in the Internet revolves around the concept of *backbones*. There is, in fact, little doubt that the Internet has a set of backbone links that carries the traffic for a majority of source–destination pairs. In other words, the traffic is not evenly spread over the various links but canalized

---

[16] See the discussion in Chapter 3.

through the more central links and hubs of the Internet. The hierarchical structure is thus manifested in two main traits (Tangmunarunkit *et al.*, 2002a). The first is that some links are more used than others. The second is that a large proportion of source–destination paths tends first to go up on the more used links and then down again on the less used ones. This amounts to an *up–down* strategy, in which a path in the Internet first finds its way up in the hierarchy, and once it has reached the backbone starts to find its way down to its final destination. This two features can be measured by studying the *load* of vertices (routers or ASs) and edges in the series of connections and vertices traversed along the paths.

The actual load of vertices and edges is not a topological quantity, and therefore cannot be directly computed from Internet maps.[17] However, it can be quantitatively estimated as the number of shortest paths among source–destination pairs that go through each edge and vertex. This definition corresponds to the vertex or edge betweenness, which indeed quantifies the level of centrality in the network. The vertex betweenness distribution shown in Figure 4.7 reveals that the load is not uniformly distributed, but has a tendency to concentrate on a few vertices, while a large number of peripheral vertices have a very small betweenness value. It is possible to identify those vertices in which most of the load is concentrated by analyzing the average betweenness $\bar{b}(k)$ of vertices with degree k, defined as

$$\bar{b}(k) = \frac{1}{NP(k)} \sum_i b_i \, \delta_{k_i,k}, \qquad (4.5)$$

where the sum is running over all vertices and $\delta_{k_i,k}$ is the Kronecker symbol with values $\delta_{i,j} = 1$ if $i = j$, and $\delta_{i,j} = 0$ if $i \neq j$. Figure 4.10 shows that $\bar{b}(k)$ is a monotonously increasing function for both the AS, AS+, and IR graphs, which indicates that the vertices carrying the largest load are indeed those with the largest degree. Those hubs, and the interconnections among them, compose a well-defined backbone of vertices and edges where the load is concentrated.[18] This conclusion is supported by the up–down analysis of shortest paths. In this case, the frequency of paths which first traverse vertices or edges strictly in order of increasing load and then of decreasing load is measured. Indeed, valid paths from a hierarchical point of view are all paths which go only down, only up, or stay flat in load values. Contrary to the up–down hierarchy, are instead all paths that have a local minimum for the load of edges or vertices within the path itself. Figure 4.11 shows the fraction of up–down paths observed in the AS and IR level, compared with

---

[17] The traffic load in the Internet is quite a subtle measure, depending strongly on the bandwidth of the physical lines and the sending patterns of individual hosts. A discussion of these issues is deferred to Chapter 10.

[18] A similar behavior is found by Tangmunarunkit *et al.* (2002a), using a more sophisticated value for the centrality, based on a minimal covering of the "traversal set," that is the set of vertex pairs whose shortest path routing traverses the edge.

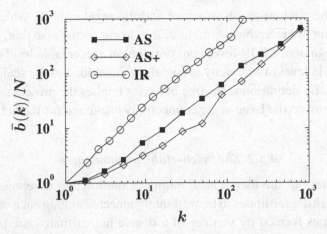

Fig. 4.10 Average betweenness as a function of the degree, $\bar{b}(k)/N$, for the AS, AS+, and IR graphs.

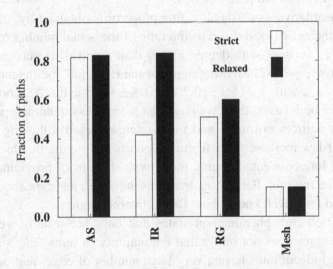

Fig. 4.11 Fraction of up–down paths for the AS and IR level maps, compared with a homogeneous random graph (RG) and a regular mesh. Data from Tangmunarunkit *et al.* (2002a).

a homogeneous random graph construction and a regular mesh (Tangmunarunkit *et al.*, 2002a). The results reported in this Figure refer to two definitions of an up–down path: *Strict* up–down paths do not allow the presence of any local minima in the load. However, *relaxed* up–down paths allow local minima, provided that the minimum load along the path is larger than the value at the end points. From Figure 4.11 we observe that the AS level map presents a larger amount of up–down hierarchy (more than 80% of the paths) than the IR map, followed as expected by the random graph structure and the regular mesh, which shows the smallest amount of hierarchical structure. It is worth noting also that, according to the relaxed definition, the IR level map presents an appreciable level of hierarchy. Indeed, a certain level of hierarchy is somehow embedded in the scale-free nature of the Internet. By definition scale-free behavior implies the presence of important hubs, which provide the large-scale connectivity backbone for the network.

4.5.2 The "rich-club" phenomenon

Another signature of the hierarchical nature of the Internet is represented by the tendency of high degree nodes to be well interconnected among each other. In other words, subgraphs formed by vertices of a degree higher than k are progressively more interconnected (Zhou and Mondragon, 2003). This implies that high degree vertices (the rich guys) form fairly well interconnected subgraphs (clubs), from which the name, *rich-club* phenomenon.

A more quantitative assessment of this property is obtained by measuring the rich-club coefficient $\phi(k)$, defined as the ratio of the actual number of edges $E_{>k}$ among the $N_{>k}$ vertices with degree higher than k and their maximum allowed number $N_{>k}(N_{>k} - 1)/2$. Measurements of the rich-club coefficient show a behavior $\phi(k) \sim k^v$, with $v = 1.1 \pm 0.2$ and 1.8 ± 0.2 for the AS and IR graphs, respectively. In both cases, thus, we have that subgraphs containing progressively higher degree vertices are more and more interconnected following approximatively a power-law increase. This feature supports the picture in which important hubs are well interconnected among them, with the aim of providing the transit backbone of the Internet. Peripherical vertices, however, just care about local connectivity and do not even know about far off Internet regions.

While the rich-club phenomenon states that hubs are usually well interconnected, this feature does not imply that the majority of hubs' edges are directed to other hubs. Indeed, hubs have a very large number of edges and only a few of them are enough to provide the connectivity to other hubs, whose number is anyway small. On average, therefore, hubs have a large majority of edges connecting to the large number of peripherical and smaller degree vertices for which they provide the connectivity to the global network. This characteristic is fully exploited by the analysis of the degree–degree correlations and the mixing properties of the graph.

4.5.3 Clustering and degree correlations

The identification of metrics able to characterize quantitatively the relation between scale-free topology and hierarchical organization has been the object of

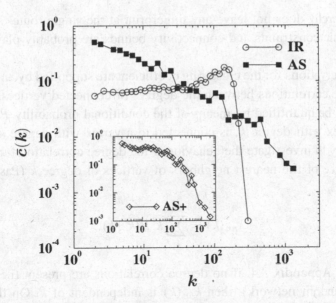

Fig. 4.12   Average clustering coefficient as a function of the vertex degree for the IR, AS, and AS+ (in the inset) maps.

several works (Tangmunarunkit *et al.*, 2002a; Vázquez *et al.*, 2002b; Ravasz and Barabási, 2003) and has prompted a detailed analysis of the clustering coefficient. In particular, important information can be gathered by inspecting the average clustering coefficient $\bar{c}(k)$ of vertices with degree k (Vázquez *et al.*, 2002b; Ravasz and Barabási, 2003)

$$\bar{c}(k) = \frac{1}{N P(k)} \sum_i c_i \, \delta_{k_i, k}, \qquad (4.6)$$

where the sum runs over all possible vertices and $\delta_{k_i, k}$ is the Kronecker symbol. As reported in Figure 4.12, in the case of the AS graph, this quantity exhibits a very clear heavy tail as a function of k, that can be approximated by a power-law decay with an exponent around 0.75. The plot corresponding to the AS+ graph shows also a noticeable heavy tail in a range of k values of three orders of magnitude; the identification with pure power-law behavior is however not clear in this case. The IR map, however, is almost constant and independent of k, with the exception of a sharp drop for large degrees, due to low statistics. These observations imply that, in the AS and AS+ graphs, vertices with a small number of connections statistically have larger local clustering coefficient than those with a large degree. This behavior is consistent with the picture of highly clustered regional networks sparsely interconnected by national backbones and international connections. However, in the IR level graph these correlations are absent. Somehow the domain hierarchy does not leave any fingerprint at the single router scale, where the geographic constraints and connectivity bounds are probably playing a more relevant role.

These observations for the clustering coefficient are supported by another metric related to the correlations between the degrees of connected vertices. These correlations can be quantified by means of the conditional probability $P(k' \mid k)$ that, given a vertex with degree k, is connected to a vertex with degree k'. A convenient quantity to investigate the behavior of the degree correlation function is the average degree of the nearest neighbors of vertices of degree k (Pastor-Satorras *et al.*, 2001)

$$\bar{k}_{nn}(k) = \sum_{k'} k' P(k' \mid k). \qquad (4.7)$$

As shown in Appendix A4, if no degree correlations are present (i.e. for an *uncorrelated* random network), then $\bar{k}_{nn}(k)$ is independent of k. On the contrary, the plots for the AS and AS+ maps exhibit a heavy tail, that can be fitted by a power-law decay for more than two decades, with a characteristic exponent close to 0.55, clearly indicating the existence of strong correlations (see Figure 4.13). This is a property referred to in physics and social sciences as *disassortative*

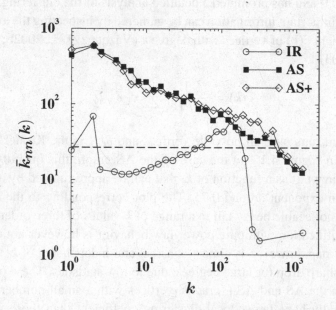

Fig. 4.13 Nearest neighbors average degree for the IR, AS, and AS+ maps. The horizontal dashed line marks the value in the absence of correlations, $\bar{k}_{nn}^{0}(k) = \langle k^2 \rangle / \langle k \rangle = 26.9$, computed for the IR map.

mixing (Newman, 2002a); i.e. high degree vertices statistically have a majority of neighbors with low degree, while the opposite holds for low degree vertices. This property is another clear signature of the structural organization of the Internet at the AS level. Vertices connectivity properties are arranged in a hierarchy of levels, in which vertices at the top levels are more interconnected with vertices at the bottom levels and vice-versa. Strikingly, there is not a finite amount of hierarchical levels. Rather we are in the presence of a continuum of levels, in that statistically each degree class has a different value $\bar{k}_{nn}(k)$. Also in this case, the IR map displays a rather different behavior. It shows a limited variation around a value very similar to that expected for an uncorrelated random network with the same degree distribution, $\bar{k}_{nn}^{0}(k) = \langle k^2 \rangle / \langle k \rangle \simeq 28$ (see Appendix A4). Noticeably, the variation has an opposite trend than in the AS level, exhibiting a mild degree of *assortative* behavior, i.e. high degree vertices tend to be neighbors of highly connected vertices. Again, the sharp drop for large k can be attributed to the low statistics for such large connectivities. Therefore, also in this case the two granularity levels show different hierarchical features.

In the statistical physics terminology, the average nearest neighbors degree $\bar{k}_{nn}(k)$ and the clustering coefficient $\bar{c}(k)$ correspond to the cumulative two- and three-point correlation functions, respectively. A natural generalization of these quantities uses higher-order metrics based on the analysis of larger *cycles*, i.e. closed paths in which all edges and vertices are distinct (Bianconi and Capocci, 2003; Caldarelli, Pastor-Satorras, and Vespignani, 2002). This kind of higher-order metrics allows the detection of ordered patterns and mesh-like structures otherwise unnoticed. For instance, the density of cycles of order 4 quantifies the statistical abundance of rectangular structures and their correlation with the vertices' degree. Noticeably, these metrics have a behavior similar to $\bar{c}(k)$, strengthening the notion of an Internet in which the hierarchical properties are extremely interwoven with the connectivity pattern. Yet it is important to stress that, at the same time, the power-law behavior of $\bar{k}_{nn}(k)$ and $\bar{c}(k)$ at the AS level is another signature of the scale-free nature of the Internet. These functions do not identify a characteristic degree value at which correlations and clustering attain a peak value. Likewise, they hint towards a continuum of hierarchies in which a wide range of correlations and clustering is allowed.

Finally, the hierarchical structure can be highlighted also by analyzing the eigenvalue spectrum of the Internet graph at the AS level. In particular, the multiplicity of eigenvalues can be related to the size of classes of subgraphs, introducing a plausible hierarchical classification of the various vertices (Vukadinovic *et al.*, 2002).

4.5.4 Hierarchical decomposition of AS graphs

The study of the inherent structural organization of the AS level has led to several hierarchical representations of the Internet. These pictures mainly provide a decomposition of the structural organization of the Internet by defining a certain number of hierarchical levels or *tiers*, usually corresponding to different connectivity classes. An early attempt by Govindan and Reddy (1997) identifies four characteristic levels on a degree scale. The lower level, the fourth one, contains all vertices with degree smaller than 4. The third level is identified by vertices with degree between 4 and 10. The second level groups all vertices with degree between 10 and 30 and the first level, the top one corresponding to national or international backbones, contains all vertices with degree larger than 30. This picture is based on a particular scale hierarchy of degree classes that is very likely not adequate for the present structure of the Internet, as clearly pointed out by the highly variable degree distribution. A different decomposition defines three hierarchy levels, *Tier-1*, *Tier-2*, and *Tier-3*, where the degree of vertices in each level is one order of magnitude larger than those of the following level (Chang, Govindan, Jamin, Shenker, and Willinger, 2001). This is a *logarithmic* grouping in classes of connectivity that might be more adequate in the case of scale-free networks.

The previous statistical analysis, however, has shown an even richer hierarchical structure. At each degree value, different statistical correlation and clustering properties are found. These properties are highly variable, with power-law behavior, defining a *continuum* of hierarchical levels, non-trivially interconnected. In particular, there is no possibility to define any degree range representing a *characteristic* hierarchical level. A schematic picture of such a hierarchical structure is obtained by combining various modular structures into each other in a recursive construction. More specifically, we can think of small groups of vertices organized in larger groups that on their turn act as the modules for the next level grouping and so on. A naïve example of such a construction is shown in Figure 4.14, in which a deterministic hierarchical network is generated by iteratively replicating an initial core structure. Interestingly, this kind of modular construction leads naturally to a scale-free degree distribution (Dorogovtsev, Goltsev, and Mendes, 2002; Jung, Kim, and Kahng, 2002; Ravasz and Barabási, 2003), while retaining a clustering hierarchy similar to that observed in the AS maps. Indeed, at the bottom level of the hierarchy we find highly clustered modules, while at the top level hubs connect different modules with low clustering, thus obtaining a clustering coefficient anti-correlated with respect to the degree.

While the above construction is far from realistic for the Internet and must be just considered as a simple conceptual example, the modular blocks from which the Internet structures are built can be schematically identified with international

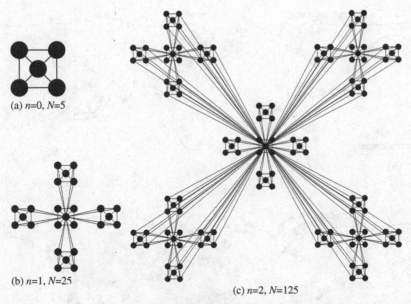

(a) $n=0$, $N=5$

(b) $n=1$, $N=25$

(c) $n=2$, $N=125$

Fig. 4.14   Iterative construction leading to a hierarchical network. Starting from a fully connected cluster of five vertices shown in (a) (note that, even though the edges are not visible, the diagonal vertices are connected), four identical replicas are created, connecting the peripheral vertices of each new cluster to the central vertex of the original cluster, obtaining a network of $N = 25$ vertices (b). In the next step four replicas of the obtained cluster are created, and the peripheral vertices are again connected, as shown in (c), to the central vertex of the original module, obtaining a $N = 125$ vertex network. The process can be iterated indefinitely. After Ravasz and Barabási (2003).

connections, national backbones, regional networks, and local area networks. Naturally, consistence with power-law distributions imposes several constraints to the degree and clustering at the different levels of the hierarchy. Any attempt towards a realistic modeling of the Internet must therefore take into account the interplay of both properties, by merging structural and degree-based rules into the simulation of Internet growth.

4.6 The Internet's geographical layout

In the previous sections we have focused on the properties of the Internet obtained from the topological layout of their maps. However, understanding of the fundamental mechanisms that drive the Internet's large-scale evolution cannot disregard the embedding of those topological properties in the physical world. The Internet requires extensive resource and time investment that must play a role in the

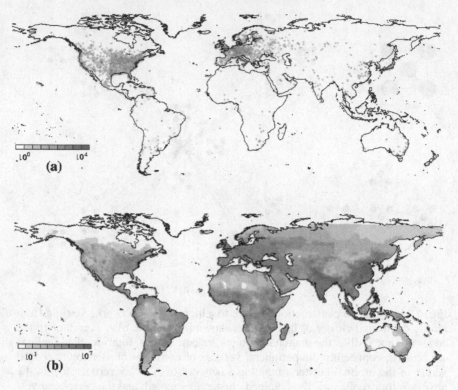

Fig. 4.15 (a) Router density map obtained by using the NetGeo tool (http://www.caida.org/tools/utilities/netgeo) to locate the geographical position of the routers mapped by the *Mercator* software (Govindan and Tangmunarunkit, 2000). (b) Population density map based on the CIESIN's population data (http://sedac.ciesin.org/plue/gpw). The resolution consists of boxes of $1° \times 1°$. The color code bars indicates increasing density values from white to black, with the highest population density of the order 10^7 people/box and the highest router density of 10^4 routers/box. After Yook *et al.* (2002).

shaping of the network, and it is crucial to correlate the Internet's topology with its geographical layout.

As a first instance, the geographical deployment of routers and the real distance among them can be inspected by identifying their geographical coordinates by using the NetGeo tool developed by CAIDA. Yook *et al.* (2002) obtained in this way a map of the world-wide router density by locating all vertices of the *Mercator* map. A visual inspection of the router density, Figure 4.15, shows an evident correlation with the population density in economically developed regions. These correlations can be more precisely quantified by studying the *box counting dimension* D_f of the density distribution (Mandelbrot, 1982). This is performed by evaluating the number $N(l)$ of boxes of size $l \times l$ with non-zero routers/inhabitants as a function of l. The analysis of a correlated region such as the USA shows

	Population	Interfaces	People per interface
Africa	837	8,379	100,011
Mexico	154	4,361	35,534
S. America	341	10,131	33,752
W. Europe	366	95,993	3,817
Japan	136	37,649	3,631
USA	299	288,048	1,061
Australia	18	18,277	975

Table 4.6 *Correlation between population (in millions)*
and number of router interfaces (Data from Lakhina,
Byers, Crovella and Matta, 2002)

that $N(l) \sim l^{-D_f}$, with $D_f = 1.5 \pm 0.1$ for both sets of data (Yook *et al.*, 2002). This implies that the router and population distributions are not homogeneous, as it would be if $D_f = 2$. The spatial distributions of both routers and population are thus fractal since they show a box counting dimension $D_f < 2$, with the very same scaling (Mandelbrot, 1982). This kind of correlation is also observed at the interface router level. Lakhina, Byers, Crovella, and Matta (2002) report on the ratio between the population and the number of interfaces in several regions of the world (see Table 4.6). From these data we conclude that the density of interfaces is strongly correlated with the wealth of the region, assuming its largest value at the most developed countries. This evidence is altogether not surprising, and shows that in technologically developed countries the Internet demand – its growth rate – is proportional to the population density.

More surprising is the analysis of the small-world properties in terms of the actual geographical distance between routers. Huffaker *et al.* (2000) have correlated the hop count with the physical distance among source–destination pairs. Interestingly, the short average hop distance of any host with respect to the rest of the network depends only very weakly on the particular vertex considered and the geographic distance among hosts. This last evidence points out that some routers are physically connected over very long distances, in order to cover with a small hop-count very large distances. This is not obvious since connecting two vertices on the Internet requires a noticeable economical investment, and network designers usually prefer to connect to the closest vertex which satisfies the bandwidth requirements. In particular, this suggests the presence of a very small fraction of long-distance connections (intercontinental cables, satellite connections) and a large predominance of shorter physical links. This fact has led to the view of an exponential decaying distance distribution on the Internet. In other words, the probability of finding two connected routers separated by a physical distance d is an exponential function, $P(d) \sim \exp(-d/d_0)$, where d_0 is a characteristic length related to geographical constraints.[19] The dependence of the router connectivity on distance is still, however, a subject of empirical debate. Focusing on the probability that two routers separated by a distance d are directly connected, Lakhina, Byers, Crovella and Matta (2002) found clear exponential behavior for several maps and continental regions, in agreement with the common view. However, Yook *et al.* (2002) studied the probability distribution of the connections of length d, finding good agreement with a power-law decay of exponent -1. Undoubtedly, this is a quite controversial issue, which deserves further experimental analysis.

All these experimental evidences must be cast in the eventual final modeling of the Internet. While in Internet topology generators these characteristics are to be imposed from the outset, a real understanding and modeling of the Internet should be able to reproduce the experimentally observed properties as the natural outcome of a few general mechanisms ruling the Internet's dynamics. The relative importance of each of these empirical finding and how much they are the manifestation of a basic force shaping the Internet are questions that will be faced in the next chapter.

---

[19] The exponentially decaying of the length of physical links has been widely used in Internet topology generators based on the Waxman model, see Section 5.3.1.

5

Modeling the Internet

At the large-scale level, the modeling of the Internet focuses on the construction of graphs that reproduce the topological properties observed in the AS and IR level maps. Representing the Internet as a graph implies ignoring the physical features of routers and connections (capacity, bandwidth, etc.), in a effort to gain a more simplified perspective that is still able to reproduce the empirical observations. From this perspective, Internet modeling initially relied on the traditional framework where complex networks with no apparent regularities were described as *static random graphs*, such as the model of Erdös and Rényi (1959). The graph model of Erdös–Rényi is the simplest conceivable one, characterized by an absolute lack of knowledge of the principles that guide the creation of connections between elements. Lacking any information, the simplest assumption one can make is to connect pairs of vertices *at random* with a given connection probability p.

Based on the random graph model paradigm, the computer science community has developed models of the Internet to test new communication protocols. The basic idea underlying the use of models to test protocols is that these should be independent (at least in principle) from the network topology. However, it turns out that their *performance* can be very sensitive to topological details (Tangmunarunkit *et al.*, 2002a; Labovitz, Ahuja, Wattenhofer, and Srinivasan, 2001; Park and Lee, 2001). The use of an inadequate model can lead to the design of protocols that run very efficiently on the model, but perform quite poorly on the real Internet. In this context, several classical *Internet topology generators* were developed, based on the Erdös–Rényi model (Waxman, 1988) or the observed hierarchical structure of the Internet (Zegura, Calvert, and Bhattacharjee, 1996; Doar, 1996). The observation of the scale-free nature of the Internet has recently prompted the construction of new generators (Jin, Chen, and Jamin, 2000; Medina, Matta, and Byers, 2000), which reproduce the correct degree distribution.

In Internet topology generators, the desired properties of the graph are imposed from the outset by designing *ad hoc* algorithms. In other words, topology generators are just a representation of our limited knowledge of the Internet. This situation is rather unsatisfactory, from a more basic point of view, when aiming to model, explain, and predict the large-scale properties and behavior of a system on the basis of the attributes and interactions of its constituent units. This strategy finds its inspiration in the statistical physics methodology, which has proved to be extremely successful in explaining the properties of matter in terms of basic elements, such as molecules and atoms, and is currently viewed as a general paradigm to bridge the gap between the local and the large-scale properties of complex systems. In the case of the Internet the ultimate goal of such an approach is to understand the observed empirical laws in terms of *emergent properties*, spontaneously developed from the microscopic dynamics of its elements.

The statistical physics approach to network modeling naturally focuses on dynamical *evolution* rules as the key elements responsible for the structural properties of the whole system. As a result of this change of perspective, a new class of models has emerged, based on the realization of two fundamental facts: (i) the Internet, as well as many other complex networks, is a *growing network*, whose number of vertices and edges continuously increases with time; (ii) connections are placed by following random processes biased by the local properties of nodes. In view of the expectations and demands of the end users, edges are established following preferential mechanisms related to the connectivity and centrality of already existing vertices. These considerations have triggered the development of degree driven growing models, whose paradigm can be found in the Barabási and Albert (1999) model. This model recovers naturally the scale-free behavior of the Internet, and from this a wealth of new models have grown up, including more sophisticated rules that take into account additional dynamical features.

In this chapter we will present a review of stochastic network models relevant to our present understanding of Internet topology, emphasizing particularly similarities to and differences from the empirical observations made on the real network. The comparison of the properties of the models with those of the Internet will help to pinpoint what are the main ingredients to be taken into account when properly modeling this complex network.

5.1 Static random graph models

The first class of models we shall consider are static, in the sense that they have a fixed size. Starting with a constant set of N disconnected vertices, these models are defined by the rules assigning edges between pairs of vertices. These models share a random nature in the process of placing the edges, that it is in general independent of the local properties of nodes. Despite this extreme simplification, however, they have provided for a long time the theoretical reference framework in network modeling, including the Internet.

5.1.1 The Erdös–Rényi model

The first theoretical model of the random network was proposed by Erdös and Rényi in the early 1960s (Erdös and Rényi, 1959, 1960, 1961). In its original formulation, the undirected graph $G_{N,E}$ is constructed starting from a set of N different vertices, which are joined by E edges whose ends are selected at random from among the N vertices. In this way, there is a total of $\binom{\frac{N(N-1)}{2}}{E}$ possible different graphs, which form a probability ensemble in that each graph has the same probability of occuring. In this respect, the previous construction resembles the *microcanonical ensemble* in classical equilibrium statistical mechanics (Pathria, 1996; Burda, Correia, and Krzywicki, 2001). A variation of this model (Gilbert, 1959) is the graph $G_{N,p}$, constructed from a set of N different vertices in which each of the $N(N-1)/2$ possible edges is present with probability p (the *connection probability*) and absent with probability $1 - p$. The relation between both constructions is straightforward: in the last case, the probability of obtaining a given graph with N vertices and E edges is

$$P(G_{N,E}) = p^E (1 - p)^{\frac{1}{2}N(N-1)-E}. \tag{5.1}$$

In the following, we will consider the Erdös–Rényi graph $G_{N,p}$, defined by the connection probability p.

Many of the properties of the Erdös–Rényi model can be easily derived. To compute the average degree, we observe that the average number of edges generated in the construction of the graph is $\langle E \rangle = \frac{1}{2}N(N-1)p$. Since each edge contributes to the degree of two vertices

$$\langle k \rangle = \frac{2\langle E \rangle}{N} = (N-1)p \simeq Np, \tag{5.2}$$

a general formula which is valid for large N. From the previous equation we observe that the average degree is diverging with the number of vertices in the graph. Since real-world graphs are most often characterized by a constant average degree, in many cases it is a natural choice to consider the behavior of the model for a wiring probability that decreases with N; i.e. $p(N) = \langle k \rangle / N$. The average degree of the random graph is also a determinant parameter in establishing the connectivity structure of the resulting network. If $\langle k \rangle < 1$ the network is composed of many small subgraphs not connected among them. For $\langle k \rangle > 1$, a giant component emerges, with size proportional to the number of vertices in the network. A more detailed account of the component structure of random graphs can be found in Chapter 6.

In order to obtain the degree distribution $P(k)$, we notice that, in a graph with wiring probability p, the probability to create a vertex of degree k is equal to the probability that it is connected to k other vertices and not connected to the remaining $N - 1 - k$ vertices. Since the establishment of each edge is an independent event, this probability is simply given by the binomial distribution

$$P(k) = \binom{N-1}{k} p^k (1-p)^{N-1-k}. \tag{5.3}$$

In the limit of large N and for $pN = \langle k \rangle$ constant, the binomial distribution can be approximated by the Poisson distribution (Gnedenko, 1962)

$$P(k) = e^{-\langle k \rangle} \frac{\langle k \rangle^k}{k!}, \tag{5.4}$$

recovering the result obtained from more rigorous arguments by Bollobás (1981). The most characteristic trait of the degree distribution of the Erdös–Rényi model is that it decays faster than *exponentially* for large k, allowing only very small degree fluctuations. The Erdös–Rényi model represents, in this sense, the prototypical example of an *homogeneous* random graph, in which, for the purpose of the large-scale characterization of the heterogeneity of the network, the degree of the different vertices can be approximately considered as constant and equal to the average degree, $k \simeq \langle k \rangle$. In Figure 5.1 we compare the analytical prediction

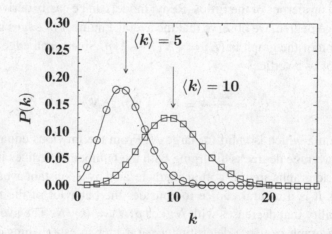

Fig. 5.1 Numerical estimate of the degree distribution of the Erdös–Rényi model for $N = 5{,}000$ and $p = 0.001$, average degree $\langle k \rangle = 5$ (circles), and $N = 10{,}000$ and $p = 0.001$, $\langle k \rangle = 10$ (squares), averaged over 100 different realizations, compared with the predicted Poisson distribution, Eq. (5.4) (full lines).

Eq. (5.4) with the result of numerical simulations of the Erdös–Rényi model. As we observe, even for moderately large sizes ($N = 5000$), $P(k)$ is very well approximated by the Poisson distribution.

The clustering coefficient $\langle c \rangle$ of the Erdös–Rényi model follows from the independence of the connections. For any vertex, the probability that any two of its neighbors are also connected to each other is given by the connection probability p. Therefore the clustering coefficient is equal to

$$\langle c \rangle = p = \frac{\langle k \rangle}{N}. \tag{5.5}$$

From this result we conclude that the clustering coefficient of the Erdös–Rényi model, at fixed $\langle k \rangle$, *decreases* with the graph size, tending to zero in the limit of an infinitely large network.

Finally, we provide some intuitions about the the the scaling of the average shortest path length ℓ with the network size N, for graphs with $\langle k \rangle > 1$, i.e. in the presence of a giant component (Newman, 2003). For a connected network of average degree $\langle k \rangle$, the number of neighbors at distance 1 of any vertex i is $\langle k \rangle$. If the position of the edges is completely random and neglecting the effect of cycles, the number of neighbors at a distance d can be approximated by $\langle k \rangle^d$. Let us define the *radius* of the graph, r_G, as the distance such that $\langle k \rangle^{r_G} \simeq N$. For $\langle k \rangle$ larger then 1, the quantity $\langle k \rangle^d$ grows exponentially fast with d, which means that an overwhelming majority of vertices are at a distance of the order r_G from the vertex i. Thus we can approximate r_G by the average shortest path length . From the definition of the radius, we obtain[1]

$$\langle \ell \rangle \simeq \frac{\log N}{\log \langle k \rangle}. \tag{5.6}$$

This approximate estimate can be proved rigorously (Bollobás, 1981), showing that the Erdös–Rényi model exhibits an average shortest path length $\langle \ell \rangle$ much smaller than the size of the graph; i.e. $\langle \ell \rangle / N \to 0$, for $N \to \infty$. This small value of the diameter is the signature of the *small-world* effect observed in many complex networks.

Despite the presence of the small-world properties, the Erdös–Rényi graph cannot be considered by any means a good representation of the physical Internet. First, it fails to reproduce many of the most characteristic properties of the Internet such as the heavy tailed degree distribution and the associated large degree fluctuations. Moreover, the mismatch in the associated clustering coefficient is also very large. At the AS level, for example (Chapter 4) the observed clustering

---

[1] It must be noted that for the full statistical ensamble of graphs with tree-like structures the average shortest path length scales as a power of N in view of the presence of linear branches (Krzywicki, 2001).

coefficient is about 0.30. An Erdös–Rényi graph with comparable size and average degree would have a clustering coefficient $\sim 4 \times 10^{-4}$, close to three orders of magnitude smaller.

5.1.2 Generalized random graphs

It is possible to extend the Erdös–Rényi model in order to construct generalized random graphs with a predefined degree distribution – not necessarily Poisson – that are otherwise random in the assignment of the edges' end-points. This procedure,[2] first proposed by Bender and Canfield (1978) and later developed in several works (Molloy and Reed, 1995, 1998; Aiello, Chung, and Lu, 2001) consists in assigning to the graph a fixed degree sequence $\{\tilde{k}_i\}$, $i = 1, \ldots, N$, such that the ith vertex has degree $\tilde{k}_i$, and afterwards distributing the end-points of the edges among the vertices, according to their respective degrees. In practice, a set of random numbers $\{\tilde{k}_i\}$ is generated and assigned to vertices according to the the selected degree distribution $P(k)$, such that $m \leq \tilde{k}_i \leq N$, with the additional constraint that the sum $\sum_i \tilde{k}_i$ must be even. The graph is completed by randomly connecting the vertices with $\sum_i \tilde{k}_i/2$ edges, respecting the degree assigned to each vertex.[3] As in the case of the Erdös–Rényi model, it is important to keep in mind that each set $\{\tilde{k}_i\}$ generates a graph realization that is a member of the ensemble of generalized random graphs with the corresponding degree distribution. As for the Erdös–Rényi model, each element of the ensemble has the same statistical weight.[4]

With the exception of the imposed constraint of having a fixed degree distribution, the graphs generated this way are in all respects random.[5] In particular, this randomness is apparent in the lack of degree correlations that allows simple analytical estimates of the graph's average shortest path length and clustering coefficient. Following Newman (2003), let us consider a generalized random graph with arbitrary degree distribution $P(k)$. Since edges are assigned at random between pairs of vertices, the probability that any edge points to a vertex of degree k is given by $kP(k)/\langle k \rangle$ (see Appendix A4). Consider now a vertex i; following the edges emanating from it, we can arrive at k_i other vertices. The probability distribution that any of the neighboring vertices has k edges pointing to other vertices different

---

[2] See Newman, Strogatz, and Watts (2001) and Newman (2003) for an introduction to the subject with applications to complex networks.

[3] As a practical recipe (Callaway, Hopcroft, Kleinberg, Newman, and Strogatz, 2001), it is useful to create a list of $\sum_i \tilde{k}_i$ elements, containing $\tilde{k}_i$ copies of the ith vertex, and join randomly chosen pairs of elements in the list.

[4] At least as long as we keep the degree sequence $\{\tilde{k}_i\}$ fixed (Newman, 2003).

[5] However, the very random nature of the vertex matching implies the necessary presence of some loops and multiple edges, which means that generalized random graphs are more properly *multigraphs* (see Appendix A1).

from i (plus the edge from which we arrived) is given by the function

$$q(k) = \frac{(k+1)P(k+1)}{\langle k \rangle}. \tag{5.7}$$

In other words, $q(k)$ gives the probability distribution of the *second nearest neighbors* that can be reached following a given edge in a vertex. The average number of these second nearest neighbors is then given by

$$\sum_k k q(k) = \frac{1}{\langle k \rangle} \sum_k k(k+1)P(k+1) = \frac{\langle k^2 \rangle - \langle k \rangle}{\langle k \rangle}. \tag{5.8}$$

The absence of correlations also yields that any vertex j is connected to another vertex l with probability $k_j k_l / \langle k \rangle N$. Notice that this expression approximates the full combinatorial form when $k_l k_j$ is larger than $\langle k \rangle N$. Thus, the clustering coefficient is simply defined as the average of this quantity over the distribution of all possible neighbors of i, i.e.

$$\langle c \rangle = \frac{1}{N \langle k \rangle} \sum_{k_j} \sum_{k_l} k_j k_l q(k_j) q(k_l) = \frac{1}{N} \frac{(\langle k^2 \rangle - \langle k \rangle)^2}{\langle k \rangle^3}. \tag{5.9}$$

It is important, however, to stress that the above equation fails for scale-free networks with too small a degree exponent. Indeed, if we consider a scale-free graph with degree distribution $P(k) \sim k^{-\gamma}$, we have that $\langle k \rangle$ is finite, while the degree fluctuations scale as $\langle k^2 \rangle \simeq k_c^{3-\gamma}$, where k_c is the maximum degree present in the graph. The value of k_c is generally related to the number of vertices in the graph (see Appendix A5) by the relation $k_c \simeq N^{1/(\gamma-1)}$. By plugging this relation in Eq. (5.9) we obtain an average clustering coefficient depending on the network size as

$$\langle c \rangle_N \simeq N^{(7-3\gamma)/(\gamma-1)}. \tag{5.10}$$

Since the clustering coefficient cannot be larger than 1, Eq. (5.9) must be restricted to values of the degree exponent $\gamma > 7/3$.

In order to provide an approximate expression for the scaling of the diameter of the graph we can proceed along the same line of reasoning used for the Erdös–Rényi model. From Eq. (5.8) we can compute iteratively the average number of neighbors z_n at a distance n away from a given vertex as

$$z_n = \frac{\langle k^2 \rangle - \langle k \rangle}{\langle k \rangle} z_{n-1}. \tag{5.11}$$

Finally, by considering that $z_1 = \langle k \rangle$, it is possible to obtain the explicit expression

$$z_n = \left(\frac{\langle k^2 \rangle - \langle k \rangle}{\langle k \rangle} \right)^{n-1} \langle k \rangle. \tag{5.12}$$

If the radius of the graph is r_G, then we must have that the number of neighbors at this distance must be approximately equal to the size of the graph N, thus obtaining $z_{r_G} = N$. Using the same argument as in the Erdös–Rényi model, we readily obtain

$$\langle \ell \rangle \approx 1 + \frac{\log[N/\langle k \rangle]}{\log[(\langle k^2 \rangle - \langle k \rangle)/\langle k \rangle]}. \tag{5.13}$$

The small-world properties are thus evident also for the generalized random graph and it is easy to check that for the case of a Poisson distribution, with second moment $\langle k^2 \rangle = \langle k \rangle + \langle k \rangle^2$, one recovers the results for the Erdös–Rényi model. However, we must keep in mind that the previous expression is only a rather crude approximation, which might fail in the presence of correlations.

In summary, in generalized random graphs the values of $\langle c \rangle$ and $\langle \ell \rangle$ depend essentially on the first and second moments of $P(k)$. For a bounded degree distribution, in which the degree fluctuations $\langle k^2 \rangle$ are finite, we observe that $\langle c \rangle \sim 1/N$ and $\langle \ell \rangle \sim \log N$, in agreement with the results found for the Erdös–Rényi model. However, for degree distributions with a fat tail, such as a power-law, the second moment $\langle k^2 \rangle$ can be very large, and even diverge with N. In this case, the prefactor in Eq. (5.9) can be noticeably large, and yield a clustering coefficient that might be higher than the one corresponding to an Erdös–Rényi graph with the same size and average degree. From this point of view, a generalized random graph reproducing the empirically observed degree distribution can be taken as a very first approximation to the topological structure of scale-free networks (Newman, 2003).

5.2 The Watts–Strogratz model

In the random graph model, the clustering coefficient is implicitly determined by the imposed degree distribution, and it is vanishing in the case of very large graphs. The empirical observation of a very large and stationary clustering coefficient in many real world networks makes it extremely interesting to find a graph construction in which it is possible to tune $\langle c \rangle$ to any desired value. Inspired by the fact that many social networks (Milgram, 1967; Wasserman and Faust, 1994) are highly clustered, while at the same time they exhibit a small average distance between vertices, Watts and Strogatz (1998) proposed a model that interpolates between ordered lattices (which have a large clustering) and purely random networks (which possess a small average path length).

The Watts and Strogatz model starts with a ring of N vertices in which each vertex is symmetrically connected to its $2m$ nearest neighbors (m on the clockwise and m in the counterclockwise sense as shown in Figure 5.2). Then, for every vertex, each edge connected to a clockwise neighbor is rewired with probability p,

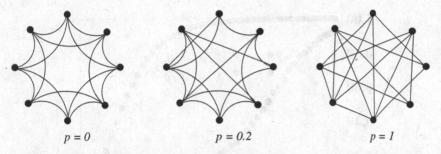

$p = 0$ $p = 0.2$ $p = 1$

Fig. 5.2 Construction leading to the Watts–Strogatz model. We start with $N = 8$ nodes, each one connected to its four nearest neighbors. By increasing p, an increasing number of edges is rewired. Rewired edges are represented as straight arcs. At $p = 1$ all edges have been rewired. After Watts and Strogatz (1998).

and preserved with probability $1 - p$. The rewiring connects the edge endpoint to a randomly chosen vertex, avoiding self-connections. The parameter p therefore tunes the level of randomness present in the graph, keeping the number of edges constant. With this construction, after the rewiring process, a graph with average degree $\langle k \rangle = 2m$ is obtained. It is however worth noticing that even in the limit $p \to 1$, since each vertex has a minimum degree m, the network retains some memory of the generating procedure and it is not locally equivalent to an Erdös–Rényi graph.

The degree distribution of the Watts–Strogatz model can be computed analytically (Barrat and Weigt, 2000), obtaining

$$P(k) = \sum_{n=0}^{\min(k-m,m)} \binom{m}{n} (1-p)^n p^{m-n} \frac{(pm)^{k-m-n}}{(k-m-n)!} e^{-pm}, \quad \text{for } k \geq m. \quad (5.14)$$

In the limit of $p \to 1$ the above expression reduces to

$$P(k) = \frac{m^{k-m}}{(k-m)!} e^{-m}, \quad (5.15)$$

a Poisson distribution for the variable $k' = k - m$, with average value $\langle k' \rangle = m$.

While the degree distribution has essentially the same features of an homogeneous random graph, the effects of the parameter p are more acute on the clustering coefficient and the average shortest path length. When $p = 0$ the number of connections among the neighbors of each node is $3m(m - 1)/2$, while the total possible number of connections is $2m(2m - 1)/2$. This yields a clustering coefficient $\langle c \rangle = 3(m - 1)/2(2m - 1)$. At the same time, the shortest path length scales as in a regular grid; i.e. $\langle \ell \rangle \sim N$. This picture changes dramatically as soon as the rewiring probability is switched on. For very small p the resulting network has a full memory of a regular lattice and consequently a high $\langle c \rangle$. In particular, Barrat

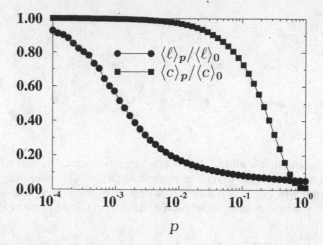

Fig. 5.3 Normalized clustering coefficient $\langle c \rangle_p / \langle c \rangle_0$ (squares) and average shortest path length $\langle \ell \rangle_p / \langle \ell \rangle_0$ (circles) as a function of the rewiring probability p for the Watts–Strogatz model. The results correspond to networks of size $N = 1,000$ and average degree $\langle k \rangle = 10$, averaged over 1,000 different realizations.

and Weigt (2000) derived the dependence of the clustering coefficient defined as the fraction of transitive triples (see Appendix A1), obtaining

$$\langle c \rangle_p \simeq \frac{3m(m-1)}{2m(2m-1)}(1-p)^3. \qquad (5.16)$$

However, even in small amounts, the edge rewiring adds *shortcuts* between distant vertices in the lattice, reducing dramatically the average shortest path length. For $p \to 1$ the network eventually becomes a randomized graph, with a logarithmically small $\langle \ell \rangle$ and a vanishing clustering coefficient. Watts and Strogatz (1998) focused on the transition between these two regimes (see Figure 5.3), noticing that, in a wide range of $p \ll 1$, the shortest path length, after decreasing abruptly, reaches almost the value corresponding to a random graph, while the clustering coefficient remains constant and equal to that of the original ordered lattice. Therefore, there is a broad region of the parameter space in which it is possible to find graphs with a large $\langle c \rangle$ and a small $\langle \ell \rangle$, as observed in most natural networks.

Interestingly, the smallest value of p at which the small-world behavior sets in is related to the size of the network (Barthélémy and Amaral, 1999b; Barthélémy and Amaral, 1999a; Barrat and Weigt, 2000). This can be qualitatively understood by noticing that the average size of the regions with a shortcut is given by the total number of vertices, N, divided by the average number of shortcuts present in the graph, pN. If the characteristic size of this region ($\sim 1/p$) is much smaller than the size of the graph, we have that short-cuts connect distant regions on the ring, producing the desired small-world effect. This immediately tells us that if

$p \gg 1/N$ the average shortest path is going to be very small. Therefore, it is in the interval $N^{-1} \ll p \ll 1$ in which the small-world properties live along with a high clustering coefficient. Noticeably, in the case of very large graphs, $N \to \infty$, even a small amount of randomness is able to produce the small-world effect.

The Watts–Strogatz model represents an important development in the modeling of social networks and many other systems (Strogatz, 2001) since it allows to associate a tuning of the clustering coefficients within the framework of static random graph theory. In addition it can explain the high clustering coefficients observed in real networks as the memory of an initial ordered structure that has been reshaped by some stochastic element. However, in the context of the Internet, the Watts–Strogatz model still misses several important features from the empirical observations. For instance, it displays a Poisson degree distribution, signaling that some other shaping principles, different from the rewiring, are missing in the definition of the model.

5.3 Internet topology generators

As has been stressed in the introduction to this chapter, the design of good Internet models is a major issue for the developing and testing of new communication and routing protocols. For this purpose, the computer science community has developed a series of *Internet topology generators*, which are used to test new technologies before actually implementing them on the Internet. All these topology generators are based on static random graph modeling, and attempt to reproduce the Internet structure with *ad hoc* algorithms tailored on the basis of the structural properties that in each case are considered as the most relevant.

5.3.1 The Waxman generator

The first Internet topology generator extensively used for protocol testing was proposed by Waxman (1988) as a simple network model on which to test the probable performance of routing algorithms. The two basic insights underlying this topology generator are:

(1) The routers in the Internet have a geographical position in space, some being far apart while others are very close to each other.
(2) Connections between distant routers are less probable than between nearby routers, due to simple economical constraints.

These two intuitions can be translated into a simple model, inspired by the Erdös–Rényi random graph, and defined by the following rules: a number N of vertices are distributed at random on a square surface of side L. Then, every pair

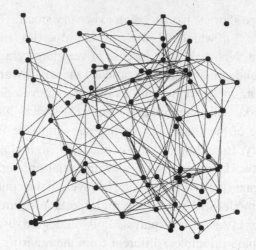

Fig. 5.4   Typical realization of the Waxman generator for $L = 100$, $N = 100$, $\alpha = 0.2$, and $\beta = 0.3$. Note the presence of isolated vertices.

of vertices i and j is considered, and joined with an edge with probability

$$P_{\mathrm{W}}(i, j) = \beta \exp\left(-\frac{d_{\mathrm{E}}(i, j)}{\alpha L}\right), \qquad (5.17)$$

where $d_{\mathrm{E}}(i, j)$ is the Euclidean distance between vertices i and j, and α and β are two real numbers in the range $0 < \alpha, \beta \leq 1$. The role of the different parameters is clear: L is the maximum distance between any two nodes, β controls the average degree of the network (larger values of β imply a larger number of edges), while α tunes the ratio between short distance and long distance edges. Figure 5.4 represents a typical network created with the Waxman generator, placing 100 vertices in a square of side $L = 100$. The parameters used are $\alpha = 0.2$ and $\beta = 0.3$.

In spite of the wide use in past years of the Waxman generator and a number of its variations (Doar and Leslie, 1993; Wei and Estrin, 1994; Zegura, Calvert, and Donahoo, 1997), it suffers from several drawbacks that render it unsuitable for a proper modeling of the Internet. Apart from the fact that the graphs generated do not have the visual aspect of graphical representations of actual Internet maps, the networks are not usually connected (see Figure 5.4); therefore, in the process of generating a correct topology, one must run the algorithm several times, checking the network for connectedness and discarding the disconnected samples. Once more, however, a most visible discrepancy with respect to the empirical observation resides in the exponentially bounded degree distribution, akin to the outcome of the Erdös–Rényi random graph model in which it is inspired, Figure 5.5. This fact renders the Waxman generator a very crude approximation to the Internet topology as we know it today.

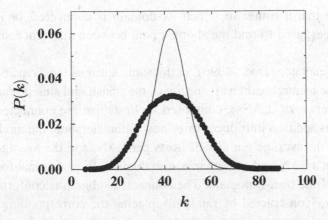

Fig. 5.5 Degree distribution for the network generated with the Waxman generator (circles) for the values $\alpha = 0.2$, $\beta = 0.3$, $L = 100$, and $N = 1,000$, averaged over 500 different realizations, compared with the degree distribution of an Erdös–Rényi random graph with the same size and average degree (full line).

5.3.2 Structural generators

A number of *structural* topology generators (Tangmunarunkit, Govindan, Jamin, Shenker, and Willinger, 2002b) have been proposed with the aim of explicitly capturing the hierarchical structure of the Internet (see Section 3.1). Among them, the two most used ones are the Transit-Stub and the Tiers generators.

The Transit-Stub (Zegura *et al.*, 1996; Calvert, Doar, and Zegura, 1997) is a topology generator that considers the two basic hierarchical elements present in the Internet, the transit and the stub autonomous systems (or *domains*). The relation between these two different levels is governed by two sets of parameters, controlling the sizes of the domains, and the connectivity between domains. The first set of parameters defines the number of transit and stub domains, and the average number of routers per transit and per stub domain. The second set specifies the average number of edges between routers in the same domain, and between routers belonging to different domains, both at the same and at different hierarchical levels.

The generation of a topology sample by the Transit-Stub generator starts from the transit level, downwards to the stub level. In the first place, each transit domain is assigned a region in space, in which its routers are distributed and joined by the corresponding number of edges, ensuring that the subgraph created this way is connected. Afterwards, the same process is followed for the stub domains. Each stub domain is then connected to one or more transit domain, and, finally, an appropriate number of stub-to-stub edges is added between randomly chosen routers in two different stub domains. On top of the connectivity relations, additional information can be included in the graph, such as pointers in stub routers indicating the transit router to which its domain is connected, or routing policies in the edges, used to find the shortest path between different routers (Calvert *et al.*, 1997).

The Tiers generator (Doar, 1996), in the same spirit as the Transit-Stub model, considers three hierarchical levels (or *tiers*), the transit and stub domains, plus the Local Area Networks (LANs), composed by hosts that are connected to the stub domains.[6] This addition introduces three new parameters into the model: the number of LANs, the average number of hosts per LAN, and the average number of edges between a LAN and routers in a stub domain. In its original formulation, it only considers one transit domain. The connected subgraphs conforming the two upper levels are constructed by randomly placing the corresponding routers in a certain region of space and joining them in a minimal spanning tree.[7] In order to reach the prescribed intra-domain connectivity, additional edges are placed from every router to the closest routers, following an order in increasing Euclidean distance. The edges between the transit and stub domains are placed by selecting a set of routers in the higher level and connecting them to a randomly chosen router in each stub domain. In order to increase inter-domain connectivity to the desired level, additional edges are placed between other routers in the stub domain and routers in the transit domain, following an increasing order in Euclidean distance from the originally connected vertex in the higher level. LANs are created with a star topology, connecting the central router to a stub domain in the same fashion as stub domains are connected to the transit domain. Figure 5.6 represents a typical sample of the graphs obtained with Tiers.

Using similar conceptual schemes, both Transit-Stub and Tiers topology generators represent a step forward in the characterization of the Internet's topology with respect to the extremely simplistic Waxman representation. These generators focus more on the detailed properties ruling the interconnectivity of different domain levels, and represent a much more realistic approximation to the topology of the network. Yet, structural generators are built with a precise average degree that leads to degree distributions with a well-defined characteristic degree and does not captures the large-scale behavior of the degree correlations of the actual Internet. However, as pointed out by Tangmunarunkit *et al.* (2002b), the degree distribution and correlations of very small networks ($\sim$100 vertices) are rather questionable concepts. Therefore, for the purpose of performing simulations of small size networks, a structural generator, with its simplified topology representation, might be a very reasonable option.

---

[6] In the original model (Doar, 1996) the transit domains are referred to as Wide Area Networks (WANs) and the stub domains as Metropolitan Area Networks (MANs).
[7] The construction of a spanning tree is a key feature of Tiers, inspired by the layout of actual networks.

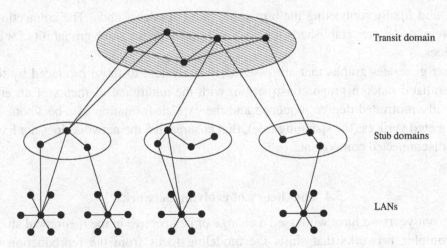

Fig. 5.6 Typical output of the Tiers Internet topology generator.

5.3.3 The Inet topology generator

After the discovery of the scale-free nature of the Internet by Faloutsos *et al.*
(1999), it became obvious that topology generators with an exponentially bounded
degree distribution were doomed to fail as faithful representations of the Internet
topology at very large scales. One of the first attempts to develop a topology generator combining some of the insights of structural generators with a power-law
degree distribution was due to Jamin, Jin, Jin, Raz, Shavitt, and Zhang (2000).
The *Inet* topology generator aims at reproducing the AS level by using the tunable degree distribution offered by the generalized random graph model. In its
latest version (Jin *et al.*, 2000), it combines empirical observations on the Internet growth with this kind of construction, to yield a random graph with a
power-law degree distribution adjusted to the form experimentally observed. The
basic parameters of *Inet* are the number of vertices of the graph N, and the fraction of vertices, p, with degree equal to 1. Assuming an exponential growth in
time for the number of vertices in the Internet maps, *Inet* extrapolates the number of months since November 1997 needed to attain the desired network size.
Then, from the empirically observed dependence on both time and degree of the
frequency on vertices of degree k and the rank[8] r of vertices of degree k, it generates a degree sequence for $(1 - p)N$ vertice,[9] that is distributed according to the
extrapolated degree distribution of the Internet at the required time. The construction of the network proceeds by creating a spanning tree with all the vertices with
degree larger than or equal to 2, attaching the vertices of degree 1 to the spanning

---

[8] The rank of a vertex is defined as the place it occupies in a list following a decreasing degree order.
[9] The rest of the vertices have degree 1 by definition.

tree, and finally connecting the remaining pairs of edges' ends. The connections between nodes are established on a random basis, with the same probability for all vertices.

Inet generates graphs that are essentially equivalent to those produced by the generalized random graph construction, with the additional elements of an empirically motivated degree sequence and the explicit formation of a backbone of connected vertices (the spanning tree), that ensures that the network does not have any disconnected component.

5.4 The theory of evolving networks

In recent years we have witnessed a change of perspective in the theoretical study of complex networks that shifts the modeling focus from the reproduction of the network's *structure* to the modeling of its *evolution*. This new approach is the outcome of the realization that most complex networks – the Internet being one of the most important examples – are the result of a growth process. The key ingredient of this new paradigm consists in considering the network as the result of the subsequent addition of new vertices and edges following a prescribed set of dynamical rules. In other words, the emphasis is on the evolutionary mechanisms that generate the observed topological properties, which become a byproduct of the system's dynamics. This methodology is akin to the statistical physics approach to complex phenomena that aims to predict the large-scale emergent properties of a system by studying the collective dynamics of its constituents.

Following the introduction of the first growing network model (Barabási and Albert, 1999), a wealth of different models have been proposed, both for undirected and directed networks, aiming to understand different aspects of real networks. In order to study these models, several analytical and numerical techniques have been borrowed from statistical physics and adapted to this new context (Dorogovtsev, Mendes, and Samukhin, 2000; Krapivsky, Redner, and Leyvraz, 2000; Dorogovtsev and Mendes, 2003). All of them are centered on solving the basic dynamical equations governing the network's formation and consider as the natural time scale for the network's evolution its size N. In this way, time is measured with respect to the number of vertices added to the graph, resulting in the definition $t = N - m_0$, m_0 being the size of the initial core of vertices from which the growth process starts. Therefore, each time step corresponds to the addition of a new vertex that establishes a number of connections (edges) with already existing vertices following a given set of dynamical rules. A full description of the system is achieved through the probability $p(k, s, t)$ that a vertex introduced at time s has degree k at the time $t \geq s$. Once known the probability $p(k, s, t)$, we can obtain the average degree $k_s(t)$ of the sth vertex at time t as

$$k_s(t) = \sum_{k=0}^{\infty} k p(k, s, t),$$
(5.18)

and the degree distribution at time t (i.e. for a network of size $N = t + m_0$) using the expression

$$P(k, t) = \frac{1}{t + m_0} \sum_{s=0}^{t} p(k, s, t).$$
(5.19)

The stationary degree distribution $P(k)$ is obtained as the very large size limit of $P(k, t)$, i.e. $P(k) = \lim_{t \to \infty} P(k, t)$. In the above discrete formulation, the evolution of the system is defined by means of a master equation (Gardiner, 1985) for the time evolution of the probability $p(k, s, t)$, that takes into account the dynamical rules adopted for the addition of edges (Dorogovtsev and Mendes, 2003).

A simpler approach that allows to clearly pinpoint the core of the dynamical evolution of growing network models is the more intuitive *continuous k approximation* (Barabási and Albert, 1999; Barabási, Albert, and Jeong, 1999; Dorogovtsev and Mendes, 2002). This approach focuses on the average value $k_s(t)$ of the sth vertex at time t, and for the sake of analytical simplicity considers the degree k and the time t as continuous variables. The properties of the system can thus be obtained by studying the dynamical rate equation governing the evolution of $k_s(t)$. This equation can be formally obtained by considering that the degree growth rate of the sth vertex will increase proportionally to the probability $\Pi[k_s(t)]$ that an edge is attached to it. In the simple case that edges are only coming from the newborn vertices, the rate equation reads

$$\frac{\partial k_s(t)}{\partial t} = m \Pi[k_s(t)],$$
(5.20)

where the proportionality factor m indicates the number of edges emanating from every new vertex. This equation is to be solved constrained by the boundary condition $k_s(s) = m$, meaning that, at the time of its introduction, all vertices have degree m. In this formulation all the dynamical information is contained in the probability $\Pi[k_s(t)]$. The properties of each model are defined by the explicit form of $\Pi[k_s(t)]$ that, as we shall see in the following sections, can accommodate also more complicated wiring processes, such as edge removal, rewiring, and inheritance. $\Pi[k_s(t)]$ represents thus the mathematical formalization of the growth rules based on optimization, expectation, or other local or global principles ruling the networks' evolution.

The dynamical evolution of the network determines also the degree distribution that can be calculated from the solution of Eq. (5.20). By considering only the

average degree $k_s(t)$, the continuous approximation smoothes off all the fluctuations and is actually equivalent to the *ansatz* solution $p(k, s, t) = \delta(k - k_s(t))$,
where $\delta(x)$ is the Dirac delta function. Inserting this expression into Eq. (5.19)
and considering the continuum limit yields

$$P(k, t) = \frac{1}{t + m_0} \int_0^t \delta(k - k_s(t)) \, ds \equiv -\frac{1}{t + m_0} \left(\frac{\partial k_s(t)}{\partial s} \right)^{-1} \Bigg|_{s=s(k,t)}, \quad (5.21)$$

where $s(k, t)$ is the solution of the implicit equation $k = k_s(t)$. We shall see in the
following sections that other properties of evolving networks are determined by the
probability rate $\Pi[k_s(t)]$. In other words, within this approximation, the form of
$\Pi[k_s(t)]$ amounts to a complete definition of the corresponding evolving network
model.

In a pedagogical perspective, it is instructive to see the above theoretical approach at work in the possibly simplest growing network model. This corresponds
to a sort of dynamical generalization of the random graph model in which each
new vertex is connected to m randomly chosen old vertices; i.e. the probability
that an existing node receives an edge is the same for all vertices (Barabási, Albert
and Jeong, 1999). By starting with a core of m_0 existing nodes, it is straightforward
to obtain

$$\Pi[k_s(t)] = \frac{1}{m_0 + t}. \quad (5.22)$$

This form of the rate probability allows the solution of Eqs. (5.20) with the boundary condition $k_s(s) = m$, yielding

$$k_s(t) = m \left[1 + \ln \left(\frac{m_0 + t}{m_0 + s} \right) \right]. \quad (5.23)$$

For large t and s, we observe that the average degree of the sth vertex increases
with t as

$$\frac{k_s(t)}{m} \simeq \ln \left(\frac{t}{s} \right). \quad (5.24)$$

This implies that the oldest vertices (with the smaller s) tend to have a larger degree, which is not surprising since they were introduced into the system first and
have had the largest opportunities to gather connections. The rate of growth of
these vertices, however, is very small (logarithmic), a fact that will induce only
small degree differences among vertices even at long times. The degree distribution, as given by Eq. (5.21), is

$$P(k, t) = \frac{1}{m(t + m_0)} [m_0 + s(k, t)], \quad (5.25)$$

and the explicit form of $s(k,t)$ is obtained inverting the relation $k = k_s(t)$ from Eq. (5.23), namely

$$s(k,t) = (m_0 + t)\, e^{1-k/m} - m_0. \tag{5.26}$$

From this result, the degree distribution is independent of time, with the following form

$$P(k) = \frac{e}{m}\, e^{-k/m}. \tag{5.27}$$

Therefore, a growing network in which edges connect new vertices with old vertices selected at random generates a homogeneous random network with an *exponentially decaying* degree distribution. This exponential behavior has been checked by means of numerical simulations of the model (Barabási *et al.*, 1999).

A final remark concerns the *physical* evolution time, T, of a growing network. As we said previously, for the sake of analytical convenience, the time scale t is measured as the number of vertices added to the network. Therefore, if we want to express the behavior of any metric as a function of the physical time T, we have to substitute t with the physical growth velocity. For instance, in the Internet the number of vertices is growing exponentially and thus we have that $t \simeq \exp(T)$.[10]

5.5 The Barabási–Albert class of models

So far we have considered models in which vertices connect to each other in a random fashion and independently from their properties. In this respect the *preferential attachment* (or *rich-get-richer*) paradigm put forward for the first time by Barabási and Albert (1999) represents a turning point in our modern view of complex networks. The insight behind this concept is the realization that, in most real networks, new edges are not placed at random but tend to connect to vertices which already have a large degree. For example, a new Internet provider probably will not connect its server with the nearest router available, but instead will want to establish a connection with a well-connected router, enabling its costumers to reach the largest possible number of servers in the minimum number of steps and with the largest bandwidth. In other words, each newborn vertex pursues the objective of obtaining a better connectivity to the network through its neighbors, despite that various factors add a probabilistic effect that not always allows the better choice. The conclusion of these observations is that new edges are not connected uniformly, but have a tendency (a larger probability) to be

---

[10] A different and more complex situation is faced in the case of the so-called *accelerated networks* (Dorogovtsev and Mendes, 2001b). In this case it is assumed that the number of new edges established in the system depends on the size of the network itself. In particular, edges may grow faster than vertices. The growth rate has the more complicated form $\Pi[k_s(t), t]$, with an explicit dependence on the growth time t.

connected with vertices that already have a large number of connections (a large degree).

Barabási and Albert (1999) combined the preferential attachment condition with the growing nature of many networks by defining a simple class of models based on the following two rules:

Growth: The network starts with a small core of m_0 connected vertices. Every time step we add a new vertex, with m edges ($m < m_0$) connected to old vertices in the system.

Preferential attachment: The new edges are connected to the old sth vertex with a probability proportional to its degree k_s.

These rules can be implemented in an algorithm that, starting from a connected initial core, generates connected graphs with fixed average degree $\langle k \rangle = 2m$ (Barabási and Albert, 1999; Barabási *et al.*, 1999). The model dynamics can be easily implemented in computer simulations and Figure 5.7 represents a typical graph of size $N = 200$ and average degree 6 ($m = 3$) generated using this algorithm. Strikingly, the numerical simulations indicate that the generated graphs spontaneously evolve into a stationary power-law degree distribution with the form $P(k) \sim k^{-3}$ (see Figure 5.8). This evidence indicates the preferential attachment

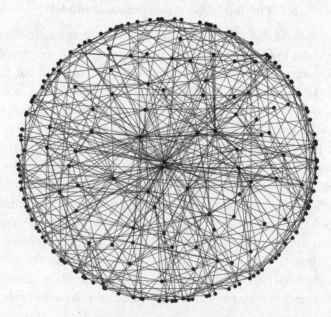

Fig. 5.7 Typical Barabási–Albert network of size $N = 200$ and average degree $\langle k \rangle = 6$. Higher degree nodes are at the center of the graph. The figure is generated with the Pajek package for large network analysis, http://vlado.fmf.uni-lj.si/pub/networks/pajek/.

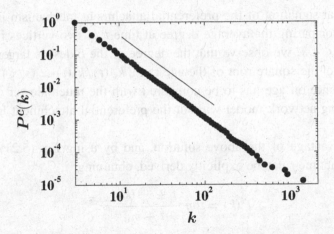

Fig. 5.8 Cumulative degree distribution of a Barabási–Albert network of size $N = 15,000$ with average degree $\langle k \rangle = 6$ in double logarithmic scale. The continuous line has slope -2, corresponding to a power-law behavior $P(k) \sim k^{-\gamma}$ with exponent $\gamma \simeq 3$.

mechanism as the dynamical driving principle that might be at the origin of heavy tailed distribution in a wide range of growing networks.

In order to gain a deeper understanding of the Barabási–Albert class of models it is possible to apply the theoretical framework described in the previous section. Within the continuous k approximation, the Barabási–Albert model is identified by the explicit form of the growth rate $\Pi[k_s(t)]$. The preferential attachment mechanism can be easily cast in mathematical form, since it states that the probability of the vertex s to acquire a new edge is proportional to its degree, obtaining

$$\Pi[k_s(t)] = \frac{k_s(t)}{\sum_j k_j(t)}, \tag{5.28}$$

where the denominator is the required normalization factor; i.e. the sum of all the degrees of the vertices in the network. Since each new edge contributes with a factor 2 to the total degree, and at time t we have added tm edges, the rate equation for k_s, Eq. (5.20), takes the form

$$\frac{\partial k_s(t)}{\partial t} = \frac{mk_s(t)}{2mt + 2m_0\langle k \rangle_0}, \tag{5.29}$$

where $\langle k \rangle_0$ is the average connectivity of the initial core of m_0 vertices. This differential equation with the boundary condition $k_s(s) = m$ can be readily solved, yielding in the limit of large networks $(t, s \gg m_0\langle k \rangle_0)$

$$k_s(t) \simeq m\left(\frac{t}{s}\right)^{1/2}. \tag{5.30}$$

This is a clear signature of the preferential attachment mechanism introduced in the model: comparing the average degree at time t of two vertices introduced at times s and $s' < s$, we observe that the degree of the oldest is larger by a factor of the order of the square root of the age ratio, $k_{s'}(t)/k_s(t) \simeq (s/s')^{1/2}$. This degree dependence on age has to be compared with the much slower one obtained in the growing network model without the preferential attachment introduced in Section 5.4.

Taking advantage of the above solution, and by using Eq. (5.21), the degree distribution at time t can be explicitly derived, obtaining

$$P(k) = 2m^2 \frac{t + \frac{m_0}{m} \langle k \rangle_0}{t + m_0} k^{-3}. \tag{5.31}$$

In the limit $t \to \infty$ we obtain the time-independent solution (valid for any $\langle k \rangle_0$)

$$P(k, t) = 2m^2 k^{-3}, \tag{5.32}$$

which indicates that the preferential attachment spontaneously generates a network with a power-law degree behavior.

The power-law distribution implicitly states that in the Barabási–Albert graph there is a non-negligible probability of finding vertices with a very large degree. These vertices act as the hubs of the network connecting a large number of nodes and thus providing connectivity shortcuts within the network. In view of this picture it is natural to expect the Barabási–Albert graph to show well-defined small-world properties. Indeed, Bollobás and Riordan (2003) and Cohen and Havlin (2003) have provided analytical evidence that the average shortest path length (and analogously the diameter) of the Barabási–Albert network scales as

$$\langle \ell \rangle \sim \frac{\log(N)}{\log \log(N)}. \tag{5.33}$$

Finally, we can address the clustering coefficient of the Barabási–Albert graph. Following Klemm and Eguíluz (2002a), let us consider the probability that the vertex appearing at time $t = s'$ establishes a connection with the sth vertex. Since edges are drawn only from the new appearing vertices, for any $s > s'$ this probability is given by the growth rate at s', obtaining

$$P(s, s') = m \frac{k_s(s')}{2ms'}, \tag{5.34}$$

where the contribution of the degree of the initial core of vertices has been neglected; i.e. $s > s' \gg 1$. The clustering coefficient of the lth vertex at time t is defined as the ratio between the average number of edges among the neighbors of l and the maximum possible number of such edges, $k_l(t)[k_l(t) - 1]/2$

(Section A1.3). Thus, in the continuous approximation, we can write

$$c_l(t) = \frac{1}{k_l(t)[k_l(t) - 1]} \int_1^t ds \int_1^t ds' \, P(l, s) P(s, s') P(s', l). \qquad (5.35)$$

Approximating $k_l(t)[k_l(t) - 1] \simeq k_l^2(t) = m^2 t/l$, the previous integral becomes independent on l, and coincides with the average clustering coefficient, yielding

$$\langle c \rangle_N = \frac{m}{8N} (\ln N)^2, \qquad (5.36)$$

where we have considered that for large networks $t \simeq N$. We thus obtain that $\langle c \rangle_N$ decreases with the network size as N^{-1} in the same fashion as the Erdős–Rényi random graph (differing only on a subleading logarithmic correction), and vanishes in the limit of infinite network size.

The interest raised by the Barabási–Albert algorithm resides in its capacity to generate graphs with a power-law degree distribution and small-world properties from very simple dynamical rules. Other features, however, such as the clustering coefficient or the fact that older nodes are always the most connected ones, do not match what we observe in the Internet. This is no wonder, of course, since Internet growth has surely other ingredients that are not considered in the Barabási–Albert algorithm, and a realistic modeling of the Internet is well beyond its scope. The very importance of the model is at the conceptual level, since it introduces a simple paradigm that is already enough to spontaneously generate highly non-trivial topological properties.

5.6 Preferential attachment revisited

The preferential attachment embedded in the definition of the Barabási–Albert model leads to a ready extension of the concept beyond the simple linear proportionality stated in the growth rate with respect to the vertices' degree. In general, one might think of a general preferential attachment rule expressed by the form $\Pi[k_s(t)] \sim k_s(t)^\alpha$, where $\alpha \geq 0$ is a fixed constant. In order to provide some general results of the effect of varying α on the resulting topology, Krapivsky *et al.* (2000) (see also Krapivsky and Redner, 2002) considered a growing network model with the rate equation

$$\frac{\partial k_s(t)}{\partial t} = m \frac{k_s(t)^\alpha}{\sum_j k_j(t)^\alpha}. \qquad (5.37)$$

In the case $\alpha < 1$, it is possible to use the *ansatz* $\sum_j k_j(t)^\alpha \equiv \mu t$ and find the parametric solution of the rate equation as

$$k_s(t) = \left[m^{1-\alpha} + \frac{m(1-\alpha)}{\mu} \ln\left(\frac{t}{s}\right) \right]^{1/(1-\alpha)}, \tag{5.38}$$

where the usual boundary condition $k_s(s) = m$ is used. From this solution, by using the Eq. (5.21) along the lines of the calculation shown in Section 5.4 , the parametric solution of the degree distribution reads

$$P(k) = \frac{\mu}{m} k^{-\alpha} \exp\left\{ -\frac{\mu}{m(1-\alpha)} [k^{1-\alpha} - m^{1-\alpha}] \right\}, \tag{5.39}$$

Finally, an explicit solution is obtained by imposing the consistency relation

$$\mu = \frac{1}{t} \sum_j k_j(t)^\alpha \stackrel{t \to \infty}{\equiv} \int_m^\infty P(k) k^\alpha \, dk, \tag{5.40}$$

where m is the minimum degree of the network. In this regime, thus, the explicit solution of the degree distribution has the form of a weak power-law with a strong stretched exponential cut-off. For $\alpha > 1$ the above method cannot be used and the general solution implies to discretize the rate equation and solve it recursively. By using this strategy it is observed that the model experiences a discrete series of connectivity transitions depending on the value of α (Krapivsky *et al.*, 2000): For $\alpha > 2$, a "winner-takes-all" situation emerges, in which all but a finite fraction of the edges are attached to a single "gel" vertex, while for $(r+1)/r < \alpha < r/(r-1)$, $r = 2, 3, \ldots$, the number of nodes with more than r edges is finite, the rest of the edges belonging to the "gel" vertex.

Noticeably, the power-law degree distribution is retained only in the case of a linear preferential attachment. The good news, however, is that at $\alpha = 1$ the power-law exponent depends on the details (additive constants, etc.) of the growth probability, leaving place for degree exponents different from the value found in the Barabási–Albert network. For instance, a linear probability of the form $\Pi(k) \sim A + k$, where A is a constant, yields a scale-free distribution $P(k) \sim k^{-\gamma}$, with a degree exponent $\gamma = 3 + A/m$, that can be tuned to any value between 3 and infinity (Dorogovtsev, Mendes and Samukhin, 2000). This opens the path up to an exploration of various modifications of the Barabási–Albert model, aiming towards a more quantitative description of the Internet graph.

From this perspective, the general heuristic result shown by Cohen and Havlin (2003) concerning the small-world property of random networks with a scale-free degree distribution assumes a particular relevance. Cohen and Havlin (2003) have presented plausible arguments that all random networks with power-law degree distributions with exponent $2 < \gamma < 3$ have very small average shortest path scaling much slower than the logarithm of N.[11] This scaling explains therefore why the average shortest path length in many networks with $\gamma < 3$ seems to be almost independent of N. For this reason it has been named *ultra small-world* behavior. This is a clear demonstration of how the presence of connectivity hubs, which are more abundant for slowly decaying power-law degree distributions, enforces a very short hop distance between vertices.

5.7 Validating the preferential attachment hypothesis

From the previous analysis, it appears that the preferential attachment rule might be the basic mechanism at the core of the heavy tailed degree distribution observed in the Internet topology. So far, however, it has been just postulated on the basis of plausibility arguments. It is therefore a priority to validate this hypothesis on empirical grounds. This can be achieved by a careful analysis of the dynamical evolution of the Internet maps. As we have seen in Chapter 4, it is possible to analyze dynamical processes for vertices and edges at the AS level by using the *Oregon route-views* maps collected by the NLANR from 1997 up to the present. For this dataset, several studies have been carried out in order to check the preferential attachment mechanism.

A direct test for the preferential attachment can be performed by measuring the functional form of $\Pi(k)$ by following its very definition (Pastor-Satorras *et al.*, 2001; Vázquez *et al.*, 2002b; Jeong, Néda, and Barabási, 2003). Different numerical techniques have been proposed to this aim. Jeong *et al.* (2003) considered the state of the Internet at a given growth time, i.e. a given size N, and recorded the number of already appeared vertices and their degree. By defining a time interval ΔT it is then possible to keep track of the relative increase $\delta k_i / \delta k$ of vertices with degree k_i, where the normalization factor δk is the total number of added edges. The ratio $\delta k_i / \delta k$ is thus an operative definition of the growth probability $\Pi(k_i)$. Furthermore, since the discrete nature of the real data yields a considerable number of fluctuations, the less noisy cumulative probability $\Pi_c(k) = \sum_{k_i=1}^{k} \Pi(k_i)$ is studied. In Figure 5.9(a), the cumulated attachment probability is plotted as a function of k showing a power-law increase, $\Pi_c(k) \sim k^{\alpha+1}$, well approximated by a value $\alpha = 1$, revealing a linear preferential attachment.

A different test considers the edges connecting newly appeared vertices in different time windows, ranging from one to three years (Pastor-Satorras *et al.*, 2001; Vázquez *et al.*, 2002b). This allows the measurement of the frequency $\chi(k)$ at which new edges are established with vertices of degree k. The frequency defines the probability that a new edge is established with anyone of the vertices with

---

[11] This result is valid as long as the graph is not a tree (see footnote in page 73).

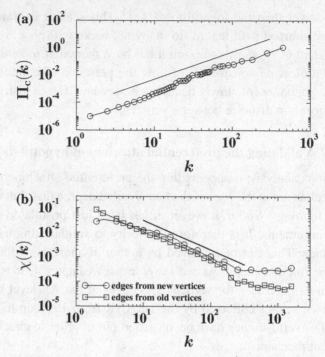

Fig. 5.9 (a) Cumulative preferential attachment for the AS level maps. The straight line has a slope 2, corresponding to a linear preferential attachment growth probability. Data provided by H. Jeong. (b) Frequency of edges emanating from new and existing vertices that attach to edges with degree k. The full line corresponds to a slope -1.2, which yields an exponent $\alpha \simeq 0.9$.

degree k. This probablity is therefore equal to the probability $\Pi(k)$ that any given vertex of degree k is selected by the new edges times the total number $NP(k)$ of vertices with that degree in a network of size N. By using the preferential attachment hypothesis $\Pi(k) = k^{\alpha}/\sum_{k} k^{\alpha}$, the frequency $\chi(k) = k^{\alpha} P(k) N/\sum_{k} k^{\alpha}$ is obtained. In Section 5.6 it has been shown that, for $0 < \alpha < 1$, in a growing network model we can approximate $\sum_{k} k^{\alpha} = \mu N$, where $\mu = \sum_{k} P(k) k^{\alpha}$ is constant in the limit of large N. This finally yields $\chi(k) = \mu^{-1} k^{\alpha} P(k)$, and, since the degree distribution has a power-law behavior $P(k) \sim k^{-\gamma}$, the behavior $\chi(k) \sim k^{\alpha-\gamma}$ is expected. Figure 5.9(b) represents the results obtained for the frequency as a function of the degree, where behavior compatible with an algebraic dependence $\chi(k) \sim k^{-1.2}$ is recovered. By using the independently obtained value $\gamma \simeq 2.1$ for the AS maps, we recover a preferential attachment exponent $\alpha \simeq 0.9$, in fairly good agreement with the linear hypothesis. Interestingly, the same linear behavior is observed for the density of edges emanating from already existing nodes, providing evidence also that the establishment of connections among existing vertices follows a preferential attachment mechanism. This could be a relevant hint for the definition of models taking into account that new edges appear not only from newly appeared vertices (see Section 5.8.1).

While both tests on the $\Pi(k_i)$ clearly show a preferential attachment mechanism, it should be noted that the evaluation of the precise value of the exponent α is still affected by numerous statistical errors. In addition, as always, the real world is not ideal and fluctuations and finite size effects come into play. By closer inspection of Figure 5.9, it is possible to see that tests reveal deviations from linear preferential attachment behavior. This occurs especially at very low and very large degree values, and it is an indication that for very small providers or very large hubs the dynamics could be mediated by different mechanisms. These deviations have also been highlighted in a different preferential attachment test performed by Willinger *et al.* (2002). In this case, AS maps in the time window spanning November 1998 until May 1999 have been considered. Each time a vertex (AS)i is added to the map with m_i edges, the degree k_j of the m_i target vertices is recorded. At the same time, the addition of the m_i edges, by using a linear preferential attachment mechanism, is simulated on the real map and the degree k_j^s of the target vertices in the simulation is registered. This procedure is repeated for all the 1,000 new nodes appearing on the maps, obtaining two sets $\{k_j\}$ and $\{k_j^s\}$, one representing the target vertices' degree in the real growth process and the other the values simulated with the pure linear preferential attachment mechanism. By graphically comparing these two datasets (Willinger *et al.*, 2002; Qian *et al.*, 2002), it is possible to show that the ASs growth shows a higher density of high degree targets than those obtained by the linear preferential attachment simulation. This is in agreement with Figure 5.9, where it appears that the frequency of edges established with large degree vertices becomes rather flat and larger than the value expected for linear preferential attachment behavior.

In summary, all empirical evidences indicate the presence of a preferential attachment mechanism, though its detailed analytical form appears to be more complex than the analytical ones considered in simple models. This indicates that if we want to model the Internet on the basis of a *degree driven* dynamics, a more faithful form of $\Pi(k_i)$ – or even the addition of extra ingredients – should be considered.

5.8 Degree driven models

After the introduction of the Barabási–Albert class of models, a large number of other network models, inspired by the degree preferential attachment mechanism, have been proposed, incorporating different ingredients in order to account for a power-law degree distribution with a connectivity exponent $2 < \gamma < 3$, local geographical factors, rewiring among existing nodes, or age effects. While an exhaustive description of all the variations of the Barabási–Albert model is beyond the scope of this book,[12] we do want to review several models which encompass some factors that might be relevant when properly modeling the Internet.

5.8.1 Wiring of edges

As real data analysis has shown (see Chapter 4), the growth of the Internet is driven by the simultaneous addition and loss of vertices, as well as by the addition and loss of new edges connecting existing vertices. This fact indicates that Internet growth is regulated by complementary wiring processes overimposed by the simple addition of new vertices' edges. This is a factor given impetus by the increasing need for backup paths and available bandwidth for data transmission.

A model that takes into account some of these processes is the generalized Barabási–Albert construction (Albert and Barabási, 2000), which includes mechanisms for the rewiring of existing edges plus the addition of new edges. The generalized Barabási–Albert construction is defined as follows: starting from a core of m_0 connected vertices, each time step one of the following operations is performed:

(1) With probability q we rewire m edges. For each of them we randomly select an edge, connecting the vertices i and j. This edge is removed and replaced by a new edge, connecting the vertex j to the new vertex i', selected with probability

$$\Pi[k_i(t)] = \frac{k_i(t) + 1}{\sum_j (k_j(t) + 1)}. \tag{5.41}$$

(2) With probability p we add m new edges. For each of them, one of the edge ends is selected at random, while the other is selected with probability $\Pi[k_i(t)]$.
(3) With probability $1 - p - q$ we add a new vertex with m edges, that are connected to vertices already present with probability $\Pi[k_i(t)]$.

For this model, the writing of the rate equation for the average degree of the sth vertex must take into account the contribution of the different rewiring-addition processes. Denoting by $N(t)$ the number of edges at time t, the growth rate due to the rewiring of edges is given by

$$\frac{\partial k_s^{(1)}(t)}{\partial t} = -\frac{mq}{N(t)} + mq \frac{k_s(t) + 1}{\sum_j [k_j(t) + 1]}, \tag{5.42}$$

where the first term represents the loss of an edge end by random removal, and the second term is the degree increase due to a rewiring with probability $\Pi[k_i(t)]$. The

---

[12] See Albert and Barabási (2002) and Dorogovtsev and Mendes (2002) and references therein to get a flavor of the work done is this direction.

drawing of new edges contributes with another growth rate expressed by

$$\frac{\partial k_s^{(2)}(t)}{\partial t} = \frac{mp}{N(t)} + mp\frac{k_s(t)+1}{\sum_j[k_j(t)+1]}. \tag{5.43}$$

Here, the first term accounts for the selection of the vertex for the random addition of an edge end, and the second one stands for the addition of an edge end with the preferential probability $\Pi[k_i(t)]$. Finally we have the contribution that accounts for the degree increase due to the addition of a new vertex

$$\frac{\partial k_s^{(3)}(t)}{\partial t} = m(1-p-q)\frac{k_s(t)+1}{\sum_j[k_j(t)+1]}. \tag{5.44}$$

Neglecting the contribution from the initial core of m_0 vertices, in the limit of large t, it is possible to write the full rate equation for the evolution of $k_s(t)$ as

$$\frac{\partial k_s(t)}{\partial t} = \frac{\partial}{\partial t}\left(k_s^{(1)}(t) + k_s^{(2)}(t) + k_s^{(3)}(t)\right)$$
$$= \frac{m(p-q)}{(1-p-q)t} + \frac{m[k_s(t)+1]}{2(1-q)mt + (1-p-q)t}, \tag{5.45}$$

where it has been considered that the total number of vertices at time t is given by $N(t) = (1-p-q)t$ (since we add a new vertex with probability $1-p-q$ every time step), and the total degree grows as $\sum_j[k_j(t)+1] = 2(1-q)mt + N(t)$.[13] This equation can be solved with the boundary condition $k_s(s) = m$, yielding

$$k_s(t) = [\mathcal{A}(p,q,m)+m]\left(\frac{t}{s}\right)^{1/\mathcal{B}(p,q,m)} - \mathcal{A}(p,q,m), \tag{5.46}$$

where the following definitions have been used

$$\mathcal{A}(p,q,m) = (p-q)\left[\frac{2m(1-q)}{1-p-q}+1\right]+1, \tag{5.47}$$

$$\mathcal{B}(p,q,m) = 2(1-q) + \frac{1-p-q}{m}. \tag{5.48}$$

In order to evaluate the degree distribution we follow the strategy shown in Section 5.4, which yields

$$P(k) \sim (k+k_0)^{-\gamma}. \tag{5.49}$$

In this result the degree offset is $k_0 = \mathcal{A}(p,q,m)$ and the degree exponent has the form

$$\gamma = 1 + \mathcal{B}(p,q,m). \tag{5.50}$$

---

[13] We have neglected the contribution from the initial core of vertices.

The scaling solution of Eq. (5.49) is valid whenever the condition $\mathcal{A}(p, q, m) + m > 0$ is fulfilled (Albert and Barabási, 2000). For fixed p and m, this condition translates into the constraint

$$q < q_{max} = \min\left\{1 - p, \frac{1 - p + m}{1 + 2m}\right\}.\qquad(5.51)$$

For $q < q_{max}$ the degree distribution of the generalized Barabási–Albert model is given by Eq. (5.49). For $q > q_{max}$ the present result is not valid and it is not possible to achieve any analytical solution. Numerical simulations, however, suggest an exponential degree distribution (Albert and Barabási, 2000).

The generalized Barabási–Albert construction yields a power-law degree distribution with an exponent tunable by changing the parameters present in the dynamics. With respect to the Internet, Qian *et al.* (2002) have shown from empirical data that the rewiring probability is very small. This observation suggests that $q \ll 1$. At the same time, however, data have shown that it is important to consider that there is a non-negligible probability for edges to disappear. While this feature is not included in the generalized Barabási–Albert algorithm, Dorogovtsev and Mendes (2000) have indeed considered a model with edge disappearance that yields a power-law degree distribution with a tunable exponent.

The fluctuations of edges have been also considered in the model proposed by Goh, Kahng, and Kim (2002), who capitalized on the scenario advanced by Huberman and Adamic (1999), where the fluctuation effects arising in the process of connecting and disconnecting edges between vertices are considered essential in the network growth.[14] In this scenario, the total number of vertices increases exponentially in time as $N(t) = N(0)\exp(t)$, and it is assumed that the degree of each vertex evolves following a stochastic multiplicative process, i.e.

$$\frac{\partial k_s(t)}{\partial t} = k_s(t)\xi_s(t).\qquad(5.52)$$

The rate of growth of each vertex is supposed to fluctuate following the stochastic variable $\xi_s(t)$. This stochastic variable has average $\langle\xi_s(t)\rangle = g$, which represents the mean average growth rate, and fluctuations $\langle\xi_s(t)\xi_{s'}(t')\rangle^2 - g^2 = \sigma\delta_{tt'}\delta_{ss'}$ which specify the fluctuation level. Noticeably, the parameters g and σ can be empirically measured (Goh, *et al.*, 2002). The multiplicative growth assumes that larger degree vertices acquire more new edges, implicitly defining an effective preferential attachment mechanism. The above formulation, however, does not specify how each vertex chooses or is chosen as a target for the edge establishment; i.e. how vertices are connected among each other. In order to provide a practical

---

[14] Actually, Huberman and Adamic (1999) proposed the model in the context of the World Wide Web (see Chapter 7).

implementation of these ideas, Goh *et al.* (2002) defined a discrete stochastic automaton based on three elementary rules, applied at each time step of the network evolution:

(1) The number of vertices is increased by a constant fraction of the total number of vertices present in the previous time step. The newly added vertices are connected to one or two previously present vertices according to the usual linear preferential attachment rule. The probability with which the new vertex connects to one or two vertices might be chosen according to empirical measurements.
(2) Each vertex increases its degree by a constant factor, the new edges being connected following the linear preferential attachment rule. The constant factor is determined by the average growth factor g measured empirically.
(3) Each vertex randomly disconnects existing edges, or connects new edges, following in this last case the linear preferential attachment rule. The probability of this event is related to the variance σ of the degree fluctuations.

It should be noted that when connecting edges, the preferential attachment rule is implemented only within the subset of vertices with degree larger than that of the vertex of the last disconnection event. This implies an adaptation process in which vertices tend to disconnect from vertices with low degree and reconnect to higher degree vertices. With these elements, the model seems to exhibit several realistic ingredients and indeed recovers a degree exponent and clustering coefficient comparable with the values shown by the Internet. At the same time, it yields a degree correlation function $\bar{k}_{nn}(k)$ with a power-law form in close analogy with the behavior exhibited by real AS level maps.

5.8.2 The fittest competition

The growing network models considered so far rely on a preferential attachment probability that depends only on the degree of the vertices present in the network and their time of arrival. In this way, at all the stages of network growth, the oldest vertices are the most connected ones and have the highest chances to become even more connected. This situation is, however, somehow unrealistic in networks such as the Internet, in which there are more factors at play than the mere degree of each vertex. For instance, economic reasons can strongly bias the choice of a new Internet provider. The connection cost increases with distance and eventually imposes a preference for a nearby, medium-sized hub, instead of the largest one that could be located far away in geographical distance. Other vertices might have more technical and economical resources, establishing a larger number of connections in a shorter time. In special cases, new technological developments, or just good marketing strategies could boost the growth of a provider at the expense of older providers. In all these situations, the possibility that newly added vertices might acquire a higher degree than the older ones should be considered.

The fitness model (Bianconi and Barabási, 2001) is an attempt to enrich the growing dynamics of the preferential attachment rule by introducing a stochastic parameter associated to each vertex, the *fitness*, that embodies all the properties, other than the degree, that might influence the probability of gaining new edges. The vertices' heterogeneity is implemented in the model by assigning to each vertex a random and fixed fitness parameter η_i, drawn from the probability distribution $\rho(\eta)$, that accounts for the resources and attractiveness of the ISP or AS. The greater the fitness, the larger the probability that new edges will be established with that vertex. In practice, the model is implemented with a growth process starting from an initial core of m_0 connected vertices. The growth proceeds by adding new vertices with randomly assigned fitness, that establish m edges with vertices already present in the network. The probability that a new edge will be established with the vertex i is given by

$$\Pi(k_i) = \frac{\eta_i k_i}{\sum_j \eta_j k_j}. \qquad (5.53)$$

This probability is proportional to both degree and fitness of the vertex so that very fit vertices can grow faster than older vertices, overcoming them and becoming the new hubs, i.e. a "fittest-get-richer" mechanism on top of the usual preferential attachment rule.

In order to obtain analytical insight into this model, it is convenient to consider that the factor $\sum_j \eta_j k_j$ can be approximated by its average value over all the possible realizations of the stochastic parameter η,[15] i.e.

$$\left\langle \sum_j \eta_j k_j \right\rangle = \int \mathrm{d}\eta \rho(\eta) \eta \sum_j k_j. \qquad (5.54)$$

Furthermore, we make the *ansatz* $\langle \sum_j \eta_j k_j \rangle = Ctm$, where C is a constant, obtaining the rate equation

$$\frac{\partial k_s(t)}{\partial t} = \frac{k_s(t)\eta_s}{Ct}. \qquad (5.55)$$

The solution of this equation with the boundary condition $k_s(s) = m$ yields

$$k_s(t) = m \left(\frac{t}{s} \right)^{\eta_s/C}, \qquad (5.56)$$

---

[15] Each sample $\{\eta_1, \eta_2, \ldots, \eta_N\}$ of the fitness parameter yields a particular network with a given degree distribution. We are interested here in computing the average distribution typical of any network realization.

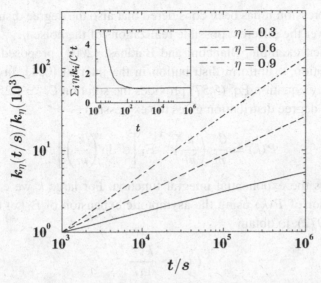

Fig. 5.10 Degree time dependence of vertices with different fitness η drawn from a uniform distribution in the interval $[0, 1]$. The inset shows the time evolution of the quantity $\sum_j \eta_j k_j$, that approaches the expected value $C^* t$. After Bianconi and Barabási (2001).

that readily implies that the degree grows as a power of time, with an exponent that depends on the specific fitness of each vertex (see Figure 5.10). Latecomer vertices with larger fitness grow more rapidly and in the long run they acquire a larger degree than their less fit predecessors.

In order to obtain an explicit expression for the degree distribution, we compute $\langle \sum_j \eta_j k_j \rangle$ as a function of C by substituting in Eq. (5.54) the solution for $k_s(t)$. This procedure yields the self-consistent relation

$$C = \int d\eta \rho(\eta) \frac{\eta}{1 - \eta/C}. \tag{5.57}$$

The right-hand term of the equation has been obtained by using the continuous k approximation and performing the limit for $t \to \infty$ (stationary case), where we have made use of the fact that $\eta/C < 1$, since the degree of any vertex cannot grow faster than t (the network size).[16] Finally, by proceeding along the general lines shown in Section 5.4, we obtain a stationary degree distribution as

$$P(k) = \frac{C}{m} \int d\eta \frac{\rho(\eta)}{\eta} \left(\frac{k}{m} \right)^{-1 - C/\eta}. \tag{5.58}$$

---

[16] Note that this condition has sense only when the range of values of the fitness parameter η is bounded.

In this last expression it has been considered that also the degree distribution must be averaged over the different possible realizations of the noise η.

As an explicit example, Bianconi and Barabási (2001) proposed the simplest fitness distribution, a uniform distribution in the interval $[0, 1]$. In this case the self-consistency equation Eq. (5.57) provides the solution $C^* \simeq 1.255$, that when inserted in the degree distribution gives the expression

$$P(k) = \frac{C^*}{m} \left(\frac{k}{m} \right)^{-1} \mathrm{E}_1 \left[C^* \ln \left(\frac{k}{m} \right) \right], \tag{5.59}$$

where $\mathrm{E}_1(x)$ is the exponential integral function. For large k we can obtain the scaling behavior of $P(k)$ using the asymptotic expansion of $\mathrm{E}_1(x)$ (Abramowitz and Stegun, 1972), to obtain

$$P(k) \sim \frac{k^{-\gamma}}{\ln k}, \tag{5.60}$$

which corresponds to a scale-free distribution of degree exponent $\gamma = 1 + C^* \simeq 2.255$, with a subleading logarithmic correction. This is a remarkable result, since assuming a uniform fitness distribution, with no special features, the model generates a network displaying a non-trivial degree distribution. In this respect, it is interesting to mention that the introduction of the vertices' fitness gives rise to a "fittest-get-richer" mechanism even in the case of static graph models as shown by Caldarelli, Capocci, De Los Rios, and Muñoz (2002).

The introduction of fitness is not the only method for taking into account the heterogeneous properties of vertices. A different perspective considers different classes of vertices exhibiting intrinsically different dynamics. This differentiation of properties evidently appears if the level of single hosts is introduced into the simulations. In this case two different classes of vertices are present, representing Internet providers and computer hosts, respectively (Capocci, Caldarelli, Marchetti, and Pietronero, 2001). Similar strategies open the road to much more articulated Internet models, with several different "actors" participating in the shaping of the network fabric (Guillaume and Latapy, 2003).

5.8.3 Technological and geographical constraints

Other important factors in a detailed large-scale modeling of the Internet are the inclusion of constraints imposed by geography and technology. As discussed in previous sections, the distance between routers is an important variable, since not all providers can afford to buy cable connections over long distances or satellite bridges. At the same time, for technical reasons, routers cannot have an unlimited number of interfaces. At the model level, the inclusion of these constraints works in a different way than the fitness model. While the inclusion of fitness introduces heterogeneity in the vertices, the constraints we are considering now act the same way on *all* vertices, imposing general properties or limitations to the growth dynamics.

A very interesting model that might represent a good abstraction for the finite capacity of routers in establishing links has been proposed by Amaral *et al.* (2000). The model evolves following the usual Barabási–Albert algorithm with linear preferential attachment, but when a vertex reaches a certain critical number of connections (a degree capacity threshold k_c) its capacity to increase its degree further is inhibited and it does not contribute further to the dynamics. While an analytical solution of this model is not available, large-scale numerical simulations show that the degree distribution has a power-law behavior that develops a cut-off at a value of k, determined by the the degree capacity k_c. This phenomenology appears particularly interesting for the IR level modeling of the Internet, since empirical data on the degree distribution show the presence of a large degree cut-off in agreement with the model.

A particular place in the modeling of the Internet is occupied by the *Brite* Internet topology generator (Medina *et al.*, 2000). *Brite* works with a fixed number of vertices; however, the edges are assigned with a preferential attachment dynamics. In addition, it considers the constraints imposed by the geographical distance among vertices. These features place *Brite* halfway between an Internet topology generator and what we have defined as a growing network model. *Brite* generates a topology on a plane divided into $L_1 \times L_1$ squares, each of which is further divided into $L_2 \times L_2$ low level squares. Each low level square can hold at most one vertex. Each high level square is assigned a number n of vertices, that can be distributed according to either a Poisson or a bounded power-law distribution. These options allow to mimick a homogeneous or a highly skewed spatial distribution of vertices, respectively. The total number of vertices is N. Each vertex is pre-assigned a random position in a different low level square. To generate the network, vertices are sequentially considered in their pre-assigned position and joined with edges to m other vertices. The assignation of edges can be *inactive*, randomly selecting one active vertex and joining it to m candidates from all the rest of the vertices; or *active*, connecting the active vertex only to nodes that have been already incorporated into the network. Finally, the vertices to be connected to the active vertex can be selected with probability proportional to the Waxman factor Eq. (5.17), proportional to the degree, as in the Barabási–Albert model, or proportional to the product of the degree times the Waxman probability. Numerical experiments (Medina *et al.*, 2000) show that with the preferential attachment element, *Brite* is able to produce a power-law degree distribution, with an exponent compatible with that found in the Internet.

One of the new elements introduced by *Brite* is the possibility of an *inhomogeneous* spatial distribution of vertices. Following this direction, Yook *et al.* (2002) put forward a topology generator in which the vertices are distributed in space, forming a scale-invariant fractal set (Mandelbrot, 1982), with a fractal dimension compatible with the value found in real router-level maps (Section 4.6). Similarly to the *Brite* generator, the probability of adding new edges is regulated by two competing mechanisms, preferential attachment and geographical distance, being directly proportional to the degree of the two vertices considered, and inversely proportional to their physical distance. Additionally, Yook *et al.* (2002) considered variations from the linear preferential attachment, different fractal dimension of the router distribution, and different forms for the attachment probability at a given distance. In this way, it is possible to study the resulting network topology in different regions of the phase space of the model's parameters. Interestingly, it appears that the topological properties observed empirically in the Internet occur only at a very particular point of the phase space, i.e. only after a careful tuning of the model's parameters. Any deviation from this point significantly alters the topological properties of the generated networks. Given the generality of this model, it is possible to find that many current generators or models lie in a particular position of this phase space, generating networks that evidently belong to different topological classes, different from the Internet. This flexible framework can thus be used as a guidance in establishing if a model based on the degree preferential attachment is a good candidate for reproducing the Internet properties.

5.9 Optimization and trade-offs

The preferential attachment paradigm defines a large category of models explaining the presence of heavy tailed degree distribution in real world graphs. However, power-law behavior is frequently observed in complex physical systems, and appears to be ubiquitous in the cooperative behavior of many social systems, ranging from the stock market (Mantegna and Stanley, 1999) to city populations (Zipf, 1949). For this reason, other generative models for power-law behavior rely more on the notion of maximizing utility and expectations of individuals. Often the interesting phenomena are related to the presence of the competition between global optimization of the system and some local constraints. In other cases it is the conflict among the various local expectations that generates non-trivial collective behavior.

An Internet model that finds its roots in this way is the *heuristically optimized trade-off* (HOT) model proposed by Fabrikant, Koutsoupias and Papadimitriou (2002). The model elaborates on the highly optimized tolerance mechanism for power-laws in designed systems devised by Carlson and Doyle (1999), and suggests that the emergence of power-laws is due to a *trade-off* mechanism; i.e.

through the optimization of the conflicting objectives pursued in the set up of the network. As a practical implementation of these ideas, the HOT Internet model of Fabrikant *et al.* (2002) is a growing model in which, at every time step, a new vertex is added to the network and placed in a random position on the unit square. The new vertex i is connected with an edge to the vertex j that minimizes the function

$$\Psi(i, j) = \alpha(N)d_{\mathrm{E}}(i, j) + \phi(j), \qquad (5.61)$$

where $d_{\mathrm{E}}(i, j)$ is the Euclidean distance between vertices i and j, $\alpha(N)$ is a constant that depends on the final size of the network, and $\phi(j)$ is a measure of the *centrality* of the vertex j. As measures of centrality, Fabrikant *et al.* (2002) propose

(1) The average shortest path length from j to the rest of the vertices in the network.
(2) The maximum shortest path length from j to any other vertex in the network.
(3) The shortest path length from j to a fixed, "central" vertex.

The edge dynamics is therefore the outcome of two conflicting objectives. The first is to limit the costs of establishing the physical connection by reducing as much as possible the Euclidean distance. The second objective is the attempt to be "centrally located" in the network, thus reducing the hop distance to other vertices. Usually, this results in a maximization of the transmission efficiency. Figure 5.11

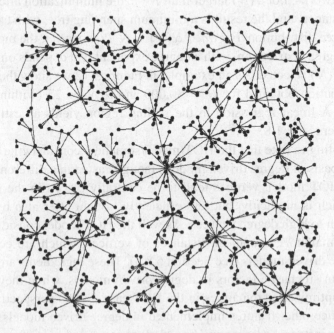

Fig. 5.11 Typical network generated with the HOT Internet model, with parameters $N = 1,000$ and $\alpha(N) = 25$. The "central" node is at the center of the box.

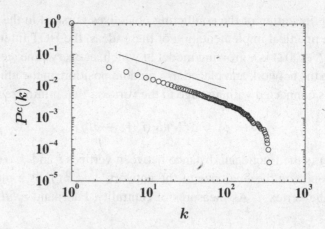

Fig. 5.12   Cumulative degree distribution for the HOT Internet model, with parameters $N = 50,000$ and $\alpha(N) = 25$. The continuous line has slope -0.8, corresponding to a degree exponent $\gamma \approx 1.8$.

shows a typical realization of this model for the last measure of centrality, generated with parameters $N = 1,000$ and $\alpha(N) = 25$.

The topology of the resulting network is found to depend on the value of the parameter $\alpha(N)$ (Fabrikant *et al.*, 2002). If $\alpha(N)$ is smaller than a constant α_0, the model is driven exclusively by the minimization of the centrality, and it generates star networks. For $\alpha(N)$ larger than $N^{1/2}$, the minimization affects only the Euclidean distance, and the result is a minimum spanning tree, with exponentially bounded degree distribution. For $\alpha_0 < \alpha(N) < N^{1/2}$, however, the model yields a power-law degree distribution, with a degree exponent that depends on the particular value of $\alpha(N)$ selected. As an example, in Figure 5.12 we show the cumulative degree distribution obtained for $N = 50,000$ and $\alpha(N) = 25$ (within the power-law regime). A linear regression in the scaling region yields an estimate of the degree exponent $\gamma \approx 1.8$.

It is interesting to note that the HOT model somehow considers the same ingredients at the basis of degree driven models, though casted in a different dynamical rule. In the HOT model, vertices tend to be centrally placed in the network, an objective which is implicitly, but less efficiently, sought after also by preferring the connection to high degree vertices. As well the HOT model considers resource constraints, smaller geographical distances of vertices, which are considered in several degree driven models (see Section 5.8.3). These ingredients are introduced as independent stochastic factors in degree driven models, while they enter in as competitive optimization dynamics in the HOT case. In this perspective, it is natural to introduce other features implemented in degree driven models in the HOT dynamics. For instance, a larger number of edges can possibly be drawn from each newly created vertex, and, more interestingly, edges can be established between existing vertices following the original trade-off's dynamics or some of its possible variations. Finally, it can also incorporate the deletion of both vertices and edges. Preliminary results in this direction show that these variations produce interesting topologies and might be used as alternative models of large-scale Internet topology (Alvarez-Hamelin and Schabanel, 2003).

5.10 Real data versus models

As we have seen in the previous sections, modern Internet topology generators (*Brite, Inet*) and growing network models based on the preferential attachment rule are able to reproduce the power-law degree distribution observed in empirical data. Models based on the HOT ideas, as well, have a wide range of parameters where heavy tailed distributions appear. The general results concerning the diameter of graphs with power-law degree distributions (see Section 5.6) imply that all these models similarly show small-world or even ultra small-world properties. With respect to these features of the Internet graphs, all these models are equally good candidates for Internet large-scale modeling. It is thus natural to wonder how these models may be validated and which parameters or metrics can be used to discriminate between good and poor models of the Internet.

At first, one would be tempted to carry out a more quantitative comparison of the degree distribution. This is, however, a slippery path. For instance, empirical data on the degree distribution do not have a precision that would allow a sharp determination of the degree exponent. As we have frequently discussed in previous chapters, the importance of the empirical evidence for heavy tailed distributions resides in their very basic differences from standard Poisson modeling. However, data do not show ideal power-law behavior and the measured exponents are subject to statistical errors, depending furthermore on the extent of the degree range considered. In summary, the measured exponents range between 1.9 and 2.4 and it would be unwise to consider the better model the one which gives rise to an exponent 2.2 instead of the one which generates the exponent 2.4. The same is true for the diameter or the average shortest path length. All measures provide values between 2 and 30, and, given the discussed logarithmic dependence on the network size, all models can be made comparably close to these values.

Once a model shows a heavy-tailed degree distribution and small-world properties, a discriminating metric can be found in the clustering coefficient. As we have seen in empirical data, IR and AS level graphs exhibit a clustering coefficient very large if compared with random graph models. Moreover, the value of the measured clustering coefficient appears to be stable along the Internet growth (see Chapter 4). The clustering coefficient is representative of the large-scale

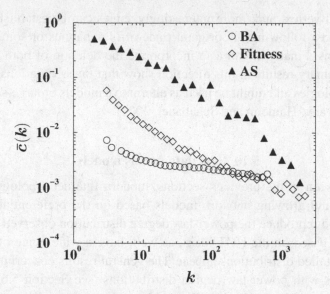

Fig. 5.13 Clustering coefficient $\bar{c}(k)$ as a function of the degree k for the Barabási–Albert and fitness networks, compared with the result from the AS map. The HOT model is not reported since in its original definition it has $c \equiv 0$.

statistical abundance of local communities and correlations. It is therefore marking a difference from all models which do not consider the appropriate level of correlations in their defining rules. For instance, despite the preferential attachment rule, the Barabási–Albert algorithm does not generate correlation between successively added vertices and thus does not have any local ordering principles (see Appendix A4). For this reason, the average clustering coefficient is vanishing when increasing the network size. The HOT model as well, in its original definition, has a null clustering coefficient, since it generates graphs with a tree-like structure.[17] In other words, both models are unable to reproduce the high clustering coefficient of Internet graphs. New ingredients must be introduced therefore to generate the opportune local cliqueness (Klemm and Eguíluz, 2002a; Alvarez-Hamelin and Schabanel, 2003).

At the AS level, the hierarchy and modular structure that have been analyzed in Chapter 4 might provide a basis for further testing and validation of models. Figures 5.13 and 5.14 represent the average clustering coefficient as a function of the degree, $\bar{c}(k)$, and the average degree of the neighbors, $\bar{k}_{nn}(k)$, respectively, for the Barabási–Albert, HOT and fitness networks. As a comparison we report the empirical data obtained from the AS level map. The figures clearly show that the

---

[17] This is due to the fact that the newly entering vertices are connected to the network by one single edge.

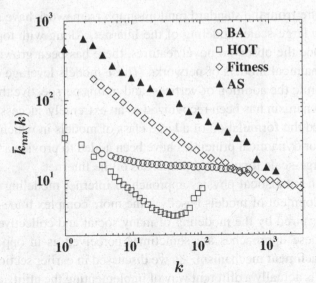

Fig. 5.14   Average degree of the nearest neighbors of a vertex $\bar{k}_{nn}(k)$ as a function of the degree k for the Barabási–Albert, HOT, and fitness networks, compared with the result from the AS map.

only model rendering results in qualitative agreement with the Internet maps is the fitness model. The lack of correlations in the Barabási–Albert and HOT networks appears to be readily emphasized by the analysis of metrics related to the hierarchical structure. These metrics, thus, appear as an interesting testbed for the validation of models aimed at characterizing the Internet at the AS level.

It is worth remembering at this point that modeling the Internet at a smaller scale level necessarily faces different issues. At the level of LANs or small internal networks, the realism assumes a much greater relevance. It then becomes very important to precisely match metrics, such as degree, the various modular structures (token rings, Ethernet, etc.), the precise layers of hierarchies among computers, etc. At this level, statistical methods loose their relevance because of the limited information that can be retrieved from the analysis of small networks. In this case, the use of detailed topology generators is needed to achieve the necessary level of realism.

5.11 The future of Internet modeling

The data collected by the various Internet mapping projects have provided empirical evidence that has drastically changed all previous views of the Internet's structure. The presence of heavy tailed distributions, complex dynamical features,

and the departure from the standard random graph framework have stimulated re-
newed effort in large-scale modeling of the Internet. Along with topology generators that include the observed novel features, there has been growing activity in
the field of dynamical models of networks. These models leverage on dynamical
principles that rule the addition of vertices, and in this perspective the preferential
attachment mechanism has been recognized as an extremely successful paradigm.
It has stimulated the formulation of a large class of models in which several additional features or dynamical principles have been added to provide a more realistic
modeling of large-scale topological properties of the Internet.

The use of the statistical physics approach to Internet modeling has also triggered the development of models based on the more complex trade-off optimization scheme, inspired by the modeling of many social and collective phenomena.
Surprisingly, these approaches are sometimes perceived as in opposition to the
preferential attachment mechanism. As we discussed in earlier sections, the trade-
off mechanism is actually a different way of implementing the utility and efficiency
principles also cast in the preferential attachment rule. It is therefore natural to see
these classes of models as inspired by a very similar philosophy, and to dwell on
the interesting combinations of the mechanisms implemented in each one of them.
For instance, one could consider the introduction of degree driven objectives in the
trade-off dynamics.

While the general tendency in growing network modeling is towards the definition of increasingly complex rules, accounting for the many dynamical features
observed in the evolution of real networks, a few major aspects have not been much
explored so far. The first consists in the local definition of the dynamical rules. As
we have discussed extensively, no one has global knowledge of the Internet's structure. It has been achieved with partial knowledge of Internet topology, and each
ISP or network administrator relies on very limited information for his connectivity or marketing choices. Connectivity, bandwidth, and centrality are variables
known only approximately, and usually on a limited scale. However, the great majority of models presuppose that each entering vertex (AS or router) might have
complete knowledge of the degree or centrality properties of the whole network.
The dynamical evolution of the network is therefore governed by *global* rules that
take into account the state of the whole network. Surely this is not the case in the
real world, and it would be very interesting to explore more deeply the effect of
introducing dynamical rules acting on a local level. Moreover, a more microscopic
formulation of Internet dynamics might also shed light on the origin of the preferential attachment mechanism that so far is just assumed on the basis of plausible
arguments and some empirical evidence.

Finally, a different aspect concerns the introduction of modularity and hierarchies in Internet modeling. While these features have been introduced *ad hoc* in

Internet topology generators, it is clear that the hierarchy represented in this way is far too simplistic with respect to real data. At the same time, the use of hierarchical constructions in growing network models is still at an early stage (Dorogovtsev *et al.*, 2002; Jung *et al.*, 2002; Ravasz and Barabási, 2003). However, empirical evidence shows that hierarchy plays an essential role in the shaping of the large-scale structure of the Internet and indicates that any realistic attempt to model this network will have to deal with these features.

6

Internet robustness

The Internet is composed by thousands of different elements – both at the hardware and software level – which are naturally susceptible to errors, malfunctioning, or other kind or failures, such as power outages, hardware problems, or software errors (Paxson, 1997; Labovitz, Ahuja, and Jahanian, 1999). Needless to say, the Internet is also subject to malicious attacks. The most common of those are the *denial-of-service* attacks, that encompass a broad set of attacks aimed at a diversity of Internet services, such as the consumption of limited resources or the physical destruction of network components (C.E.R. Team, 2001). Given so many open chances for errors and failures, it might sometimes be surprising that the Internet functions at all.

The design of a computer network resilient to local failures (either random malfunctions or intentional attacks) was indeed one of the main motivations for the original study of distributed networks by Paul Baran (1964). Considering the worst possible scenario of an enemy attack directed towards the nodes of a nationwide computer network, Baran analyzed the "survivability" (defined as the average fraction of surviving nodes capable of communication with any other surviving node) of the several network designs available at that time. His conclusion was that the optimal network, from the survivability point of view, was a mesh-like graph with a sufficient amount of redundancy in the paths between vertices.[1] Even in the case of a severe enemy strike, depleting a large number of components, such network topology, would ensure the connectivity among the surviving computers, diverting communications along the ensemble of alternative paths. In addition, this sort of resilience against attacks would be economically beneficial, since it would allow the use of low-cost, unreliable connections between components, which would not substantially hinder communications in case of a random failure.

---

[1] That is, a graph in which there is sufficient number of alternative, different paths between every pair of vertices.

Baran's proposal for a designed distributed network was finally not taken into account in the development of the primitive ARPANET, nor in the subsequent evolution of the Internet. This network has undergone instead a self-organized growth, following no pre-established plan, and its present scale-free topological structure is far from that envisioned by Baran. In this respect, the possible fragility of the Internet to random failures and intentional attacks is a major issue, with several practical implications.

The study of the resilience of the Internet to failures is not an easy task. After any router or connection fails, the Internet responds very quickly by updating the routing tables of the routers in the neighborhood of the failure point. Therefore, the error tolerance of this network is a dynamical process, which should take into account the time response of the routers to different damage configurations (see Chapter 10). Though, a first approach to the analysis of the Internet's robustness can be achieved at the *topological* level by studying the behavior of the AS and IR level maps under the removal of vertices or edges. These studies have shown that the Internet presents two faces in front of component failures: it is extremely robust to the loss of a large number of randomly selected vertices, but extremely fragile in response to a targeted attack.

In this chapter we provide a review of results on the topological resilience of the Internet to damage. We shall present numerical experiments which show that the Internet can withstand a considerable amount of random damage and still maintain overall connectivity in the surviving network. In particular the Internet's tolerance to massive random damage is much higher than for meshes or random homogeneous networks, suggesting that the cause for this robustness resides in its power law degree distribution. This intuition will find analytical confirmation by mapping the damage problem into percolation phase transitions. The very nature of Internet degree distribution, on the other hand, implies the presence of heavily connected hubs. A targeted attack, aimed at knocking down those hubs, has dramatic consequences for Internet connectivity. In this case we shall see that the deletion of a very small fraction of hubs is enough to break the network down into small, isolated components, hugely reducing its communication capabilities.

6.1 Internet robustness to random failures

A first empirical assessment of the robustness of the Internet in front of random failures can be obtained by studying the topological response of Internet maps to the removal of edges or vertices. Focusing on the effect of vertex removal, and assuming that all vertices are equally likely to experience a failure, this theoretical experiment can be performed on a connected map of size N by looking at the effect achieved by removing a fraction g of randomly selected vertices. The deletion of a vertex implies that all the edges adjacent to it are also removed. In order to monitor the response of the network to the damage, one can control several topological metrics related to network connectivity (Albert, Jeong, and Barabási, 2000a). A first and natural quantity to study is the the size S_g of the largest component in the network after damage with respect to the size S_0 of the undamaged network. In particular, a ratio $S_g/S_0 > 0$ indicates that a macroscopic fraction of vertices is still capable of communication, i.e. a giant component still exists. On the contrary, $S_g/S_0 \simeq 0$ signals that the whole network has been fragmented into small disconnected components, each one of them not containing an appreciable fraction of the vertices.[2]

A natural question to ask in this context concerns the maximum amount of damage that the network can take, i.e. the *threshold* value of the removal probability, g_c, above which $S_g/S_0 \simeq 0$ and the network can be considered destroyed. In Figure 6.1 we report the behavior of S_g/S_0 under a progressive density of damage g in the case of the IR level map. The result is compared with the behavior obtained for a regular mesh, namely the square lattice, and the Erdös–Rényi random graph that, given its Poisson degree distribution, can be considered a typical example of the homogeneous random network with $\langle k^2 \rangle \sim \langle k \rangle^2$. The figure provides striking evidence that the IR graph behaves very differently from both regular meshes and the Erdös–Rényi random graph. While at low levels of damage the resilience is essentially determined by local details, such as the minimum degree, which are not taken into account in this general discussion, at large levels of damage both the square lattice and the Erdös–Rényi random graph exhibit a threshold value g_c over which S_g/S_0 abruptly drops to zero. The IR graph, on the contrary, has a much higher tolerance to large damage, and for values of the damage as high as 85% of the total network still shows a small but macroscopic fraction of connected surviving vertices.

The evidence for a distinctive tolerance to large levels of random damage is confirmed by the analysis of the Internet at the AS granularity. In this case, along with S_g/S_0, we also report the average shortest path length of the giant component, $\langle \ell \rangle_g$, as a function of the fraction of randomly removed vertices g. Figure 6.2 shows the behavior of these two quantities in random removal experiments performed on the AS and AS+ Internet maps. For both maps the slow decrease at large g of the relative size of the largest component is confirmed. In fact, when 50% of the vertices have been removed, the largest cluster has a size close to 40% of that of the initial network, while for 80% damage the largest component still contains 10% of the original vertices. The average shortest path length , however, is slightly increasing (with a relative variation of only a few percent), up

---

[2] $S_g/S_0 = 0$ occurs only in the so-called thermodynamic limit in which $S_0 \to \infty$.

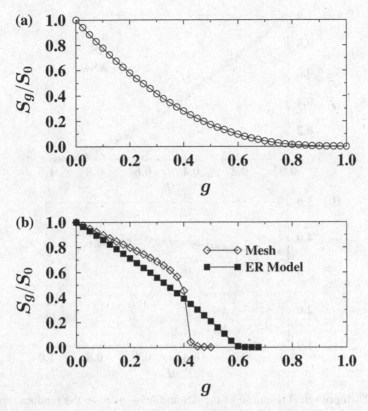

Fig. 6.1 Topological resilience of IR level map to the random removal of vertices (a), compared with a square lattice and an Erdös–Rényi random graph with the same average degree than the IR graph (b). The latter plots show a clear drop at the corresponding threshold of damaged vertices, while in the case of the IR map the decay is slower, with no apparent sign of a sharp threshold.

80% damage, above which it decreases quite rapidly. This fact seems to indicate a threshold value $g_c \approx 0.9$ below which the surviving network can function almost as efficiently as the undamaged Internet.[3]

Other metrics have been proposed in the literature to characterize the Internet's resilience to vertex damage. For example, Park, Khrabrov, Pennock, Lawrence, Giles, and Ungar (2003) remark that S_g and $\langle \ell \rangle_g$ are not representative of the overall connectivity of the surviving network, since they focus exclusively on the largest surviving component, thus neglecting the contribution of the small surviving clusters of connected vertices, different from the largest component. To take into account this contribution, they propose to measure instead two different metrics: the fraction K of all the connected pairs of vertices with respect to the total

---

[3] At least in what respect the average distance that the IP packets have to cross in order to reach a random point in the surviving network.

Internet robustness

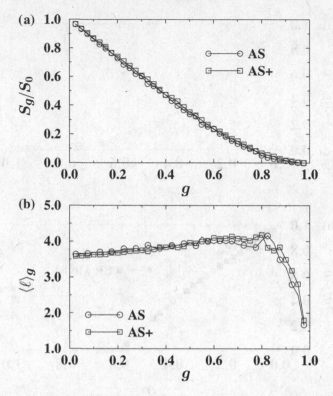

(a)

(b)

Fig. 6.2   Topological response of the AS and AS+ maps to the random removal of vertices. (a) Relative size of the largest connected cluster in the network with respect to the size of the undamaged network. (b) Average shortest path length of the largest cluster in the damaged network.

number of pairs in the network, and the average shortest path length $\langle \ell \rangle_g^c$ between *all* pairs of vertices with the shortest path length $\ell_{ij} \neq \infty$. Studying in particular the quantities K and $\langle \ell \rangle_g^c / K$, Park *et al.* (2003) were able to confirm the results concerning the considerable resilience to damage of Internet maps.

A different and interesting analysis has been performed by Crucitti, Latora, Marchiori, and Rapisarda (2003). In this work, the authors point out that previous metrics overlook the effect of lost connectivity in the fragmented network, and propose to measure the *efficiency* of the graph as the average of the inverse of the shortest path length between all pairs of vertices (Latora and Marchiori, 2001)

$$E = \frac{2}{N(N-1)} \sum_{i<j} \frac{1}{\ell_{ij}}. \tag{6.1}$$

This metric avoids the problems associated with pairs of vertices belonging to disconnected clusters, that yield a contribution $1/\ell_{ij} = 0$, while giving a measure of the traffic capacity of the whole network. Indeed, for a well-connected network,

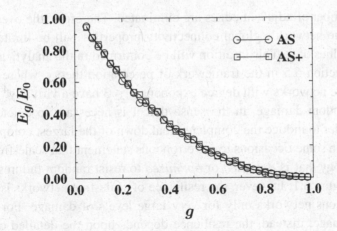

Fig. 6.3 Relative efficiency of the AS and AS+ maps as a function of the fraction of removed vertices in a random attack.

with small shortest path length, we expect to observe considerable efficiency. However, for a damaged network in which the average distance between vertices is large, we should expect a decrease in the efficiency. Figure 6.3 shows the relative efficiency E_g/E_0 measured in the AS and AS+ maps as a function of the density of random damage g. In analogy to the behavior of Figure 6.2, the efficiency of both maps slowly decreases with g, signaling the intrinsic robustness of the Internet to random damage.

The conclusion from the analysis presented in this section is that the Internet appears as very *robust* with respect to large densities of random failures at both the IR and AS levels. The network is able to sustain a considerable number of disabled components and still present an appreciable fraction of the vertices in the largest component. At the same time the average distance between vertices in the giant component is more or less constant, and approximately equal to the value in the original, undamaged network, thus keeping the topological properties of the network still beneficial for the sake of communication purposes. This is not the case for regular meshes or random networks, where a sharp damage threshold above which the network is completely fragmented appears much earlier.

Since the main topological feature of the Internet resides in its power-law degree distribution, it is natural to think of this property as the one responsible for the Internet's high resilience. Indeed, it is possible to understand this point qualitatively by recalling that the power law form of this distribution implies that the vast majority of vertices have a very small degree, while a few hubs collect a very large number of edges, providing the necessary connectivity to the whole network. When removing vertices at random, chances are that the largest fraction of deleted elements will have a very small degree. Their deletion will imply, in turn, that only a limited number of adjacent edges are eliminated. Therefore, the overall damage exerted on the network's global connectivity properties will be limited, even for very large values of g. This intuition will be confirmed in the analytical study performed in Section 6.5 in the framework of percolation theory, where it is shown that scale-free networks with degree exponent $\gamma \leq 3$ have a virtually[4] infinite tolerance to random damage, in the sense that it is necessary to delete a fraction $g \rightarrow 1$ in order to induce the complete breakdown of the largest component. This fact has led on some occasions to the erroneous statement that scale-free networks have a topology that is *designed* or *optimized* to resist random failures. As is evident from Figure 6.1, however, the resilience of scale-free networks is larger than in homogeneous networks only for very large levels of damage. For very small levels of damage, instead, the resilience depends upon the detailed connectivity properties of the network, including the minimum degree.[5]

6.2 Resilience to damage as a percolation phase transition

The natural theoretical framework to understand the resilience of the Internet to random failures is that of *percolation theory* (Bunde and Havlin, 1991; Stauffer and Aharony, 1994). Percolation, first introduced in the 1940s in the context of gelation processes, is the simplest model describing a disordered system capable of experiencing a phase transition. In order to define a *vertex* percolation process, let us consider a regular square lattice, Figure 6.4, in which each vertex is occupied with probability p, and empty with probability $1 - p$. Let us define a *cluster* as the connected network made by a set of vertices which are nearest neighbors.[6] When p is small, Figure 6.4(a), the clusters are made by a reduced number of vertices, much smaller than the total size of the lattice. However, when p is larger than a certain critical value p_c, called the *percolation threshold*, a *percolating cluster* appears, joining two opposite sides of the lattice, Figure 6.4(b). The number of vertices belonging to the percolating cluster diverges when increasing the size of the lattice; therefore it is also called the *infinite cluster*. The emergence of the infinite cluster corresponds to the onset of a giant component, which contains a finite fraction of the lattice's vertices. It is also possible to define *edge percolation* processes, in which the edges or bonds between vertices are the ones likely to be occupied with probability p. In this case, two vertices belong to the same cluster if there is path of occupied edges connecting them.

---

[4] The apparent thresholds $g_c \approx 0.9$ observed in the numerical experiments will turn out to be the effect of the finite size of networks, i.e. an effective threshold dependent on the maximum degree of the network.

[5] For instance, the mesh topology is the one that seems to perform better at very low damage levels, in agreement with the original analysis of Baran (1964).

[6] Clusters are identified by the connected set of vertices joined by black thick lines in Figure 6.4.

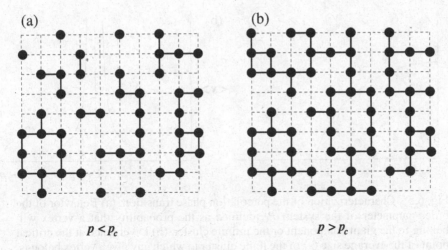

(a)   (b)

$p < p_c$ $p > p_c$

Fig. 6.4   Examples of site percolation on a two dimensional regular lattice. The filled circles represent occupied vertices. (a) System below the percolation transition. (b) System above the percolation transition. Note the presence of the infinite cluster, connecting the upper with the lower boundary.

Percolation is useful for modeling a variety of physical systems in which disorder plays a relevant role. For example, consider occupied vertices that are electrical conductors, while empty vertices are insulators, and that the electrical current is only able to flow between nearest neighbors conductors. At small p adjacent conductors form small, isolated islands, and the system is overall an insulator. When p is large, however, there is an infinite cluster of conductors connecting two opposite sides of the lattice, and the system behaves as a conductor. The insulator/conductor transition takes place abruptly, at percolation point p_c.

The threshold p_c defines a *critical* point at which the system undergoes a dramatic change in its macroscopic properties, i.e. a *phase transition*. In order to characterize quantitatively this change of properties, one can focus on the probability P_G that a randomly chosen vertex belongs to the infinite cluster (or the giant component). In infinite systems, as shown in Figure 6.5, P_G is identically zero for all values $p < p_c$. Above this threshold, the infinite cluster appears and P_G becomes an increasing function of p. The probability P_G is generally referred to as the *order* parameter, in the sense that it characterizes the onset of some macroscopic ordering or global structure in the system, and p is defined as the *critical* parameter of the system.

Percolation theory has been defined on regular lattices embedded in a D-dimensional space. In random graphs with N vertices, however, there is no embedding space and any vertex can be connected to any other vertex. This is equivalent to work in a space that is N-dimensional; i.e. any vertex has N possible neighbors.

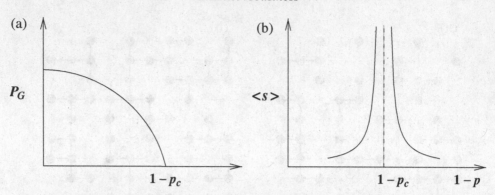

Fig. 6.5   Characterization of the percolation phase transition. (a) Behavior of the order parameter of the system P_G defined as the probability that a vertex will belong to the giant component or the infinite cluster. (b) Divergence at the critical point of the average size $\langle s \rangle$ of the finite cluster to which any given vertex belongs.

For instance, the onset of the giant component of the Erdös–Rényi random graph is analogous to an edge percolation problem in which the $N - 1$ edges of each vertex can be occupied with probability p. In the $N \to \infty$ limit, often considered in random graph theory, the problem is therefore analogous to infinite dimensional edge percolation, for which it is usually possible to provide a simple solution, as we shall see in Section 6.3.1. In this framework, the macroscopic behavior of networks, when faced with random or targeted removal of vertices or edges, finds a natural characterization in terms of an *inverse* percolation process in a random graph. In this context, the lattice in which percolation takes place is the graph under consideration. In the undamaged graph, with $g = 0$, all the vertices are occupied. The deletion of a fraction g of vertices corresponds to a random graph in which the vertices are occupied with probability $p = 1 - g$. For small g, we are in the region of p close to 1, in which the infinite cluster (identified as the giant component) is present. The threshold for the destruction of the giant component, $g_c = 1 - p_c$, can be thus computed from the percolation threshold at which the infinite cluster first emerges. In this case the phase transition corresponds to the separation of a region of damages which still allow a connected network of appreciable size from a region in which the system is totally fragmented. The order parameter P_G is a function of $g = 1 - p$ and it can be defined as $P_G = S_g/S_0$, where, as in the previous section, S_g is the size of the largest component after damage g, and S_0 is the size of the original network. However, the relation $P_G = \lim_{S_0 \to \infty} S_g/S_0$ defines unambiguously the transition point only for the infinite size limit. In the finite system, the transition is smoother and the order parameter never attains a null value above the threshold, relaxing to its minimum value $P_G \sim 1/S_0$.

The mapping between damage and inverse percolation processes results particularly useful for providing analytical insight on network topological robustness.

In the following sections we shall provide a general presentation of percolation theory and will lean on these results in order to achieve a full understanding of network resilience to damages.

6.3 Percolation theory

Percolation theory focuses on the behavior of the percolation transition close to the critical point where it takes place. A basic quantity describing the system structure is the *cluster number distribution* $n_s(p)$, defined as the number of clusters of size s per lattice vertex, at the percolation probability p. The probability that any given vertex belongs to a cluster of size s is therefore $sn_s(p)$, where the fact that the vertex can be any one of the cluster's elements has been considered. For $p < p_c$, when there is no infinite cluster, the probability p that a vertex belongs to any one of the finite clusters can be expressed as

$$\sum_s s\, n_s(p) = p, \qquad p < p_c. \tag{6.2}$$

Above the critical point p_c, the infinite cluster appears and $P_G > 0$. Since any occupied vertex belongs either to the infinite cluster or to a cluster of finite size, we can generalize Eq. (6.2) and write

$$P_G + \sum_s{}' s\, n_s(p) = p, \qquad p > p_c, \tag{6.3}$$

where the prime in Eq. (6.3) means that the giant component is excluded from the summation.

The probability that an occupied vertex belongs to a cluster of size s is given by $s\, n_s(p)/\sum_s{}' s\, n_s(p)$, where the infinite cluster has been excluded to avoid divergences. The average size $\langle s \rangle$ of the cluster to which any occupied vertex belongs is therefore given by

$$\langle s \rangle = \frac{\sum_s{}' s^2\, n_s(p)}{\sum_s{}' s\, n_s(p)}. \tag{6.4}$$

At $p < p_c$ the average cluster size $\langle s \rangle$ is finite. Above the threshold the unrestricted average size is infinite because an infinite cluster appears. This implies that $\langle s \rangle$ develops a divergence at $p = p_c$ that is absorbed in the infinite connected cluster above the threshold. The $\langle s \rangle$ is thus a singular function at p_c as shown in Figure 6.5.

The divergence of $\langle s \rangle$ contains a lot of physical information about the system and it is the fingerprint of critical phase transitions. Indeed, $\langle s \rangle$ is proportional to the second moment of $n_s(p)$, that must develop long tailed behavior in order to allow for the singularity occurring at p_c. As has been discussed at large in Chapter 4, diverging moments are typical of distributions with power-law tails. Therefore it is possible to formulate the general scaling ansatz (Stauffer and Aharony, 1994)

$$n_s(p) = s^{-\tau} f[s/s_c(p)], \qquad \text{with} \qquad s_c(p) = (p - p_c)^{-1/\sigma}, \qquad (6.5)$$

where τ and σ are exponents, whose values depend on the dimensionality and other properties of the system. The function $f(x)$ is supposed to be constant for $x \ll 1$, and rapidly decreasing for $x \gg 1$. The quantity $s_c(p)$ plays the role of a size cut-off: only connected clusters with a size smaller or comparable with $s_c(p)$ are present and define the physical properties of the system for any given value of p. The present ansatz allows a finite $\langle s \rangle$ below the critical point with an exponentially bounded $n_s(p)$. Approaching the critical point, $n_s(p)$ becomes more and more heavy tailed. Finally when $s_c(p) \to \infty$ at $p \to p_c$, the simple power-law behavior $n_s(p) \sim s^{-\tau}$ is attained with the corresponding diverging behavior of $\langle s \rangle$.

Power-law behavior and singular functions are typical of critical phase transitions, where the onset of a macroscopically ordered phase (for instance the presence of a global connected structure) is anticipated by large fluctuations in the statistical properties of the system. It is only when these fluctuations become of the order of the system size itself, at the critical point, that the macroscopic order arises and the system enters the new phase region. This overall picture is common to all critical phase transitions such as the liquid-vapor and paramagnetic-ferromagnetic transitions at the critical point (Binney *et al.*, 1992).

The power-law behavior close to p_c can be generalized to other quantities by providing a scaling theory of percolation. In analogy with other phase transitions it is possible to write

$$P_G \sim (p - p_c)^{\beta}, \qquad (6.6)$$

$$\langle s \rangle \sim (p_c - p)^{-\gamma}, \qquad (6.7)$$

valid for percolation in any kind of lattice, and in which β and γ are named *critical exponents*. Indeed, assuming the scaling form (6.5), it is possible to provide expressions for the probability P_G and the average cluster size. Substituting into Eq. (6.4) the form postulated for the cluster distribution, and approximating the summation by an integral, we obtain

$$\langle s \rangle \sim \frac{1}{p} \int s^{2-\tau} f[s/s_c(p)] \, ds \sim s_c(p)^{3-\tau} \int x^{2-\tau} f[x] \, dx. \qquad (6.8)$$

The integral in x is bounded and well defined and the only dependence on p is given by $s_c(p) = (p - p_c)^{-1/\sigma}$, yielding

$$\langle s \rangle \sim (p - p_c)^{(\tau-3)/\sigma}. \qquad (6.9)$$

From here we recover the scaling form in Eq. (6.7), with an exponent

$$\gamma = \frac{3 - \tau}{\sigma}.$$
(6.10)

To compute the behavior of P_G, it is possible to use Eq. (6.3). Noticing that we are very close to the critical point[7] $p_c \approx \sum_s s\, n_s(p_c)$, we obtain

$$P_G \approx \sum_s{}' s\,[n_s(p_c) - n_s(p)] \sim \int s^{1-\tau}\,[f[0] - f[s/s_c(p)]]\ ds$$
$$\sim s_c(p)^{2-\tau} \sim (p - p_c)^{(\tau-2)/\sigma}.$$
(6.11)

From this expression we can finally read the exponent

$$\beta = \frac{\tau - 2}{\sigma}.$$
(6.12)

Eqs (6.10) and (6.12) are scaling relations which reduce to two the number of independent exponents defined by percolation theory.

The power law behavior and the scaling theory for percolation have been extensively checked for different kinds of lattices in both analytical studies and computer simulations (Stauffer and Aharony, 1994) and it is worth mentioning the results concerning the *universality* of critical exponents. Indeed, the exact value of the critical exponents does not depend on the fine details of the percolation model. In general, they just depend on the system's dimensionality and symmetries of the order parameter. Thus, while the exact value of p_c is different in a triangular or rectangular lattice embedded in a two dimensional space, the critical exponents result to be the same in both cases. Universality emerges thus as a fundamental concept in physics, where different systems, as diverse as magnetic or liquid-vapor systems but with the same symmetry and dimensionality, are described by the same critical behavior (Binney *et al.*, 1992; Yeomans, 1992).

6.3.1 A simple example: percolation on the Cayley tree

In order to have an explicit example of the percolation theory in networks, it is possible to consider the percolation transition on a Cayley tree. This particular system has the advantage of being analytically solvable, and it is equivalent to the study of percolation in an infinite dimensional hypercubic lattice, or in the so-called mean-field limit (Stauffer and Aharony, 1994).

A Cayley tree is constructed starting from a central vertex with z edges. The number z is called the coordination number of the tree. Each edge connects to a

---

[7] Note that close to p_c we expect $P_G \approx 0$.

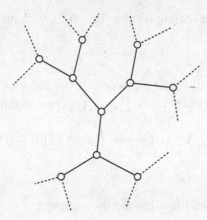

Fig. 6.6   Cayley tree with coordination number $z = 3$.

vertex, which is the origin of other $z - 1$ edges. The construction proceeds iterating this process to infinity, with $z - 1$ new edges emanating from each new vertex, with no loops nor "dangling ends" (vertices with only one edge), except in the boundary of the intermediate steps of the process.[8] Figure 6.6 represents the first stages of a Cayley tree with coordination number $z = 3$.

Taking advantage of the absence of loops and the independent probability for the placement of occupied vertices, we can easily solve vertex percolation in the Cayley tree. Following Stauffer and Aharony (1994), let us construct a infinite path of edges between occupied vertices, starting from any occupied vertex. Following any edge, we find $z - 1$ new edges, that lead on average to $p(z - 1)$ new occupied vertices. We can construct an infinite path only if there is at least one occupied vertex at the end of any of those $z - 1$ edges, i.e., if $p(z - 1) > 1$. Since an infinite path can only exist in the presence of the infinite connected cluster, this relation defines the percolation threshold

$$p_c = \frac{1}{z - 1}. \tag{6.13}$$

In order to compute the behavior of P_G, let us define q as the probability that a given edge does not lead to a vertex connected to the infinite cluster. The event that an edge does not lead to the infinite cluster is equal to the event that it leads to an empty vertex, or either that it leads to an occupied vertex whose $z - 1$ emerging edges do not lead to the infinite cluster. Since these are independent events, it is possible to write the self-consistent equation

$$q = (1 - p) + pq^{z-1}. \tag{6.14}$$

---

[8] The absence of loops is one of the arguments for the infinite dimensionality of the Cayley tree.

The solution $q = 1$ correspond to all edges leading to finite clusters, which is the correct answer for $p < p_c$. The behavior at $p > p_c$ will therefore be given by the real positive solution of Eq. (6.14) with $q < 1$. However, the probability P_G is by definition equal to the probability that a given vertex is occupied and that any of its z edges lead to the infinite cluster, i.e.

$$P_G = p(1 - q^z).\tag{6.15}$$

Eq. (6.14) cannot be solved for a general value of z. Turning to the particular case $z = 3$ (for which the percolation threshold is $p_c = 1/2$) we obtain the solution $q = (1 - p)/p$, valid for $p > 1/2$. Substituting this result into Eq. (6.15), we obtain

$$P_G = p\left[1 - \left(\frac{1-p}{p}\right)^3\right].\tag{6.16}$$

As for the average cluster size $\langle s \rangle$, let us define T as the average size of the cluster that is reached by following any edge. This edge leads to an empty vertex with probability $1 - p$, and to an occupied vertex with probability p, which contributes to the average cluster size with itself, plus the $z - 1$ edges that emanate from it. Therefore, $T = p[1 + (z - 1)T]$, whose solution is $T = p/[1 - p(z - 1)]$ for $p > 1/(z - 1)$. The average cluster size is due to the contribution of an occupied vertex plus its z outgoing edges; therefore

$$\langle s \rangle = 1 + zT = \frac{1 + p}{1 - p(z - 1)}.\tag{6.17}$$

From Eqs. (6.16) and (6.17) we can estimate the value of the critical exponents of the various scaling relations. For the particular case $z = 3$, performing a Taylor expansion in the vicinity of $p = p_c = 1/2$ gives

$$P_G \simeq 6(p - p_c),\tag{6.18}$$

$$\langle s \rangle \simeq \frac{3}{4}(p_c - p)^{-1},\tag{6.19}$$

which recover $\beta = \gamma = 1$. Finally by using the scaling relations (6.10) and (6.12) the exponents $\tau = 5/2$ and $\sigma = 1/2$ are obtained. The same critical exponents hold for any value of z. These values can also be derived rigorously (Essam, 1980), providing a direct confirmation of the validity of the scaling ansatz made in Eq. (6.5).

6.4 Percolation transition in random graphs

The percolation transition and the appearance of a giant connected component in graph theory refer with a different language to the same critical phenomenon. In network language, the critical point translates into the existence of a given threshold condition on the graph connectivity properties that marks a separation between a regime in which the network is fragmented in a myriad of small subgraphs and a regime in which there is a giant component containing a macroscopic fraction of the network's vertices. In the infinite size limit this corresponds to the onset of an infinite component containing a finite fraction of vertices (see Appendix 1).

The study of percolation in generalized random graphs finds its natural formulation in the generating functional technique (Callaway, Newman, Strogatz, and Watts, 2000; Newman, 2003). Here we will consider, however, a more approximate argument, analogous to that applied in the previous section for percolation in a Cayley tree, valid for sparse graphs (Dorogovtsev and Mendes, 2003).[9] Let us then consider a random uncorrelated graph with arbitrary degree distribution $P(k)$, and let us define q as the probability that a randomly chosen edge in the network does not lead to a vertex connected via the remaining edges to a component of infinite size. Neglecting cycles, the quantity q can be self-consistently computed as the probability that an edge leads to a vertex of degree k, times the probability that none of its $k - 1$ emanating edges is leading to an infinite component, averaged over all the possible values of k. In uncorrelated networks, since the probability that an edge is connected to a vertex of degree k is $kP(k)/\langle k \rangle$, we have that

$$q = \sum_k \frac{kP(k)}{\langle k \rangle} q^{k-1}.$$ (6.20)

The probability that a vertex does not belong to the giant component, $1 - P_G$, is equal to the probability that it has degree k and none of its edges leads to an infinite component, averaged over all the degrees k, yielding

$$P_G = 1 - \sum_k P(k)q^k.$$ (6.21)

As in the analysis of percolation in a Cayley tree, the solution $q = 1$ in Eq. (6.20) indicates the absence of a giant component. This component can only exist whenever there is a positive real solution of Eq. (6.20) with $q < 1$. In order to find this solution it is useful to resort to a geometrical argument. The solution of Eq. (6.20) is given by the intersection of the curves $y_1(q) = q$ and $y_2(q) = \sum_k kP(k)q^{k-1}/\langle k \rangle$. The function $y_2(q)$ is monotonously growing and concave with q between the limits $y_2(0) = P(1)/\langle k \rangle < 1$ and $y_2(1) = 1$. Therefore, in order to have a solution $q^* < 1$, the slope of $y_2(q)$ at $q = 1$ must be larger than or equal to 1 (see Figure 6.7). This condition translates into the equation

---

[9] That is, graphs which have a local tree structure with no cycles.

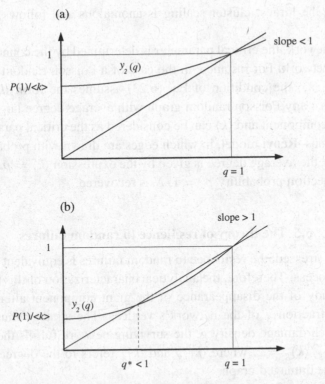

Fig. 6.7 Graphical solution of Eq. (6.20). (a) When the slope at $q = 1$ of the auxiliary function $y_2(q)$ (see text) is smaller than 1, the only solution is the trivial $q = 1$. (b) When the slope is larger than 1, the additional solution $q^* < 1$ appear.

$$\frac{\mathrm{d}}{\mathrm{d}q}\left(\sum_k \frac{kP(k)}{\langle k \rangle} q^{k-1}\right)\Bigg|_{q=1} \equiv \frac{\langle k^2 \rangle - \langle k \rangle}{\langle k \rangle} \geq 1. \qquad (6.22)$$

The equality in Eq. (6.22) signals the onset of the formation of the giant component, which happens precisely at the point $\langle k^2 \rangle - \langle k \rangle = \langle k \rangle$. This condition, which can also be written as

$$\frac{\langle k^2 \rangle}{\langle k \rangle} = 2, \qquad (6.23)$$

is exact in the case that cycles are statistically irrelevant, as it is the case for random uncorrelated graphs in the $N \to \infty$ limit close to the transition point, and was first derived on more rigorous grounds by Molloy and Reed (1995). It marks the critical point of a phase transition, separating the phase (for $\langle k^2 \rangle / \langle k \rangle < 2$) in which all the components are trees and the size of the largest component at most scales as $\ln(N)$, from the phase (for $\langle k^2 \rangle / \langle k \rangle > 2$) in which there exists a giant component scaling as N, while the size of the other components at most scales as $\ln(N)$. At the point

$\langle k^2 \rangle / \langle k \rangle = 2$, the largest cluster scaling is anomalous and follows the behavior $N^{2/3}$.

In this framework, the critical parameter is determined by the connectivity properties of the network. For instance, in the case of a Poisson random graph where $\langle k^2 \rangle = \langle k \rangle^2 + \langle k \rangle$, the condition of Eq. (6.23) assumes the form $\langle k \rangle_c = 1$. This readily tells that any Poisson random graph with average degree larger than 1 exhibits a giant component and $\langle k \rangle$ can be considered as the critical parameter. In the case of the Erdös–Rényi model, in which edges are drawn with probability p (see Section 5.1.1), the average degree is given by the expression $\langle k \rangle = pN$ and finally a critical connection probability $p_c = 1/N$ is recovered.

6.5 The theory of resilience to random failures

As previously stressed, the resilience to random failures is equivalent to an inverse percolation process. Therefore, the analytical characterization of this transition relies on the study of the disappearance of the giant component after the random removal of a fraction g of the network's vertices. The strategy thus consists in finding at which damage density g, the surviving network fulfills the percolation condition $\langle k^2 \rangle_g / \langle k \rangle_g = 2$, where $\langle k^2 \rangle_g$ and $\langle k \rangle_g$ refers to the degree distribution moments of the damaged graph.

Following the intuitive approach proposed by Cohen, Erez, ben Avraham, and Havlin (2000), let us consider a sparse uncorrelated generalized random graph with degree distribution $P_0(k)$ and first moments $\langle k \rangle_0$ and $\langle k^2 \rangle_0$. Removing a fraction g of all vertices in the graph is equivalent to removing a fraction g of the neighbors of any surviving vertex. Therefore, in the damaged network any surviving vertex with original degree k' will have a degree k with probability

$$\binom{k'}{k}(1-g)^k g^{k'-k}, \qquad k' \geq k, \tag{6.24}$$

which essentially corresponds to the random deletion of $k' - k$ neighboring vertices, preserving k of them.[10] The degree distribution of the network defined by the surviving vertices will then be given by

$$P_g(k) = \sum_{k'=k}^{\infty} P_0(k') \binom{k'}{k}(1-g)^k g^{k'-k}. \tag{6.25}$$

From this equation we can compute the first and second moments of the degree

---

[10] Note that this argument is valid for the deletion of a fraction g of either neighboring vertices or edges. The results here obtained, therefore, are equally valid for both vertex and edge percolation.

distribution in the damaged network, obtaining

$$\langle k \rangle_g = \sum_k k P_g(k) = (1 - g)\langle k \rangle_0, \tag{6.26}$$

$$\langle k^2 \rangle_g = \sum_k k^2 P_g(k) = (1 - g)^2 \langle k^2 \rangle_0 + g(1 - g)\langle k \rangle_0. \tag{6.27}$$

These expressions can be plugged in Eq. (6.23) that gives the condition for the presence of a giant component in the surviving network, yielding the precise value of the threshold g_c as the one that satisfies the equation

$$\frac{\langle k^2 \rangle_{g_c}}{\langle k \rangle_{g_c}} = \frac{(1 - g_c)\langle k^2 \rangle_0}{\langle k \rangle_0} + g_c = 2. \tag{6.28}$$

By using the heterogeneity parameter $\kappa = \langle k^2 \rangle_0 / \langle k \rangle_0$, the explicit solution for the removal threshold is obtained as

$$g_c = 1 - \frac{1}{\kappa - 1}. \tag{6.29}$$

That is, the critical threshold for the destruction of the graph's connectivity differs from unity by a term that is inversely proportional to the degree fluctuations of the undamaged network.[11] This readily implies that the topological robustness to damage is related to the graph's degree heterogeneity. In homogeneous networks, in which the heterogeneity parameter is $\kappa \simeq \langle k \rangle$, the threshold is finite and will simply depend on the average degree. In highly heterogeneous networks, in which $\kappa \gg \langle k \rangle$, the threshold approaches larger values, being dominated by the magnitude of $\langle k^2 \rangle$. In particular, all scale-free graphs with diverging $\langle k^2 \rangle_0$ have $\kappa \to \infty$, therefore exhibiting an infinite tolerance to random failures; i.e. $g_c \to 1$. Scale-free graphs thus define a class of networks characterized by a distinctive resistance to high levels of damage that is clearly welcome in many situations.

6.5.1 Removal threshold and finite-size corrections in scale-free graphs

The results shown in the previous section point at scale-free graphs as very robust networks that in principle can be broken apart only by damaging all their vertices. However, real-world networks, the Internet among them, necessarily show finite size effects due to resource or size constraints. It is therefore interesting to analyze the behavior of the critical threshold with respect to the various details characterizing the realistic degree distributions empirically observed.

---

[11] This result should be compared with the threshold obtained in epidemic processes that will be encountered in Chapter 9. The similar dependence on the degree fluctuations shown by both phenomena reveals the deep connection between epidemic spreading and percolation. This can be shown at a rigorous level by mapping the random percolation process into the susceptible-infected-removed epidemic model (Newman, 2002b).

As a first case, let us consider growing scale-free graphs made of N vertices, with degree distribution $P_0(k) = ak^{-\gamma}$, where a is a normalization constant. The finite number of vertices implies a maximum degree $k_c \leq N$ to the network so that the available degree range is $k \in [m, k_c]$, where m is the minimum degree. In this case it is possible to show from general scaling arguments that the cut-off k_c is a growing function of the graph size $k_c(N)$ (see Appendix A5). In the limit of large N ($k_c(N) \gg m$), and assuming the continuous k approximation, the parameter κ can be explicitly calculated and takes the form[12]

$$
\kappa \simeq \begin{cases} \frac{\gamma-2}{\gamma-3}m & \text{for } \gamma > 3, \\ \frac{2-\gamma}{\gamma-3}mk_c(N)^{3-\gamma} & \text{for } 2 < \gamma < 3, \\ \frac{2-\gamma}{3-\gamma}mk_c(N) & \text{for } \gamma < 2. \end{cases} \tag{6.30}
$$

For $\gamma > 3$, κ assumes a constant value independent of the network's size or cut-off. The network thus always exhibits a transition, located at the intrinsic threshold value

$$
g_c = 1 - \frac{1}{\frac{\gamma-2}{\gamma-3}m - 1}. \tag{6.31}
$$

As expected, the more interconnected is the network (the largest is m), the closer the threshold to 1, i.e. a larger amount of damage is needed to collapse the network. This result shows that, for what concerns random damage, scale-free graphs with a large degree exponent behave similarly to regular lattices or graphs with Poisson degree distribution.

The picture is drastically different for scale-free graphs with $\gamma \leq 3$. In this case, $\kappa \to \infty$ when $N \to \infty$, which means that the threshold tends to 1 in the limit of an infinite graph size. That is, scale-free graphs with degree exponent smaller than 3 are not *vulnerable* to the random deletion of vertices, since in order to eliminate the giant component it is necessary to remove almost all vertices forming the network. Graphs with a finite size, however, do exhibit a non-zero effective threshold which depends on the actual size N of the network. In particular, for $2 < \gamma < 3$ (the case of interest for the Internet graphs), the threshold reads

$$
g_c(N) \simeq 1 - \left(\frac{3-\gamma}{\gamma-2}\right)k_c(N)^{\gamma-3}. \tag{6.32}
$$

Yet this threshold is generally quite close to 1 at relatively small network sizes. For instance, by using the empirical values of the degree distribution of the AS graphs,

---

[12] In the case of $\gamma = 3$, logarithmic corrections take over in the second moment of the degree distribution and $\kappa \sim \ln k_c(N)$.

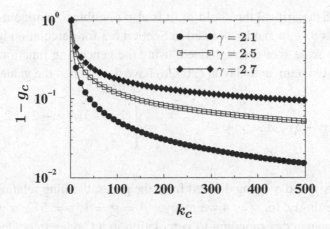

Fig. 6.8 Threshold for random removal of vertices in graphs with degree distribution $P_0(k) = ck^{-\gamma} \exp(-k/k_c)$ as a function of k_c and for different degree exponent γ. For all $\gamma \leq 3$ the network has a very high resilience with $g_c \to 1$ for $k_c \to \infty$. It also appears that networks with lower γ have higher threshold and thus a larger tolerance to damage.

the direct application of the theoretical prediction Eq. (6.29) yields $g_c \approx 0.9$. This is in reasonable agreement with the analysis of the AS and AS+ maps performed in Section 6.1.[13]

A different kind of finite-size effects are those introduced by physical or resource constraints (see Section 4.4). In this case an external cut-off k_c, not depending on the system size, determines a degree distribution of the form $P_0(k) = ck^{-\gamma} \exp(-k/k_c)$, for $k \geq 1$. Inserting this distribution into Eq. (6.29) it is possible to compute the threshold as a function of the undamaged cut-off k_c, for different values of the degree exponent (Callaway *et al.*, 2000). Figure 6.8 reports the predicted behavior of g_c, which also in this case tends to 1 for $k_c \to \infty$. In addition it is clear that the threshold approaches larger values more rapidly in networks with smaller γ. This behavior suggests a larger robustness for degree exponents closer to 2, due to the larger relative density of hubs that keep the network connected. It is interesting to note in this case that even relatively small values of the cut-off ($k_c \simeq 50$) leave almost intact the networks robustness, still yielding a threshold $g_c \approx 0.8 - 0.9$. This scenario might be relevant in the case of the IR level map, that shows a cut-off at degree $k_c \simeq 100$ (see Section 4.4). Indeed, in this case a good agreement with the theoretical prediction and the empirically observed threshold is recovered.

[13] The comparison of the theoretical results with the empirical data must be considered only as indicative. In fact, the theory refers to scale-free graphs without degree correlations. Instead, degree correlations are present in AS level graphs (see Section 4.5).

Along with the critical threshold g_c, it is also possible to work out the values of the exponents β, γ, σ, and τ defined in Section 6.3 for percolation in an uncorrelated random scale-free graph. Indeed, using the generating functional technique, Cohen, ben-Avraham, and Havlin (2002b) have determined the values

$$
\beta = \begin{cases} 1/(3-\gamma) & 2 < \gamma < 3 \\ 1/(\gamma - 3) & 3 < \gamma < 4 \,, \\ 1 & \gamma > 4 \end{cases} \qquad \tau = \begin{cases} (2\gamma - 3)/(\gamma - 2) & 2 < \gamma < 4 \\ 5/2 & \gamma > 4 \end{cases} \,,
$$

$$(6.33)$$

the exponents γ and σ being derived from the general scaling relations (6.10) and (6.12). Interestingly, for $\gamma > 4$ we recover $\beta = \gamma = 1$, $\tau = 5/2$, $\sigma = 1/2$, which are the exponents corresponding to percolation in a Cayley tree. Once more, this evidence confirms that uncorrelated random scale-free graphs with large exponents $\gamma > 4$ are equivalent with respect to percolation to homogeneous random graphs with Poisson degree distribution and infinite trees.

As a final remark, it is interesting to note the effect that the presence of degree *correlations* can have in percolation on scale-free networks (Newman, 2002a; Vázquez and Moreno, 2003). For networks in which correlations can be defined by the conditional probability $P(k'\,|\,k)$ that a vertex of degree k is adjacent to a vertex of degree k' (Appendix A4), it is possible to show that the damage threshold is given by $g_c = 1 - 1/\tilde{\Lambda}_m$, where $\tilde{\Lambda}_m$ is the largest eigenvalue of the matrix $\mathbf{C} = \{C_{kk'}\}$, with elements $C_{kk'} = (k' - 1)P(k'\,|\,k)$, provided that $\tilde{\Lambda}_m > 1$ (Vázquez and Moreno, 2003). In this case, the resilience of the network ($g_c \to 1$) depends on the divergence of the eigenvalue $\tilde{\Lambda}_m$ instead of the fluctuations of the degree distribution, as was the case of random networks. Nevertheless, it can be proved that this eigenvalue diverges for all scale-free networks with degree exponent $\gamma \leq 3$, except in the peculiar case in which the divergence is due to the vertices of degree 1 (Boguñá, Pastor-Satorras, and Vespignani, 2003b).

6.6 Internet's Achilles heel

The scale-free nature of the Internet protects it from random failures, since the hubs, that hold the network together with their many links, are difficult to hit in a random selection of vertices. Since hubs are the key elements ensuring the connectivity of the network, however, it is easy to imagine that a *targeted* attack, aimed at the destruction of the most connected vertices, should have a very disruptive effect. This possibility can be checked by analyzing the connectivity properties of Internet maps in which a fraction g of the highest degree vertices have been deleted (Albert, Jeong, and Barabási, 2000b; Broido and Claffy, 2002). In practice this is made by

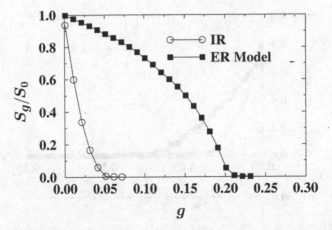

Fig. 6.9 Topological resilience to targeted attacks of the IR level Internet map and an Erdös–Rényi random graph with the same average degree. By using this damaging strategy, the IR map appears as as the most fragile network. Even a density of removal as low as the 0.05% is enough to completely fragment the network.

removing the vertices following a decreasing degree ordered list. The first vertex to be removed is therefore the one with the highest degree. Then the second highest degree vertex is removed, and so on until the total number of removed vertices represents a fraction g of the total number of vertices forming the network. As for the random removal, the behavior of the giant component measured as S_g/S_0 can be studied for increasing damages g. In Figure 6.9, the topological resilience to targeted attack of the IR level map is compared with that of an Erdös–Rényi graph with the same average degree. In this case the emerging scenario is opposite to that found for the random removal damage. Strikingly, the IR map appears much more vulnerable than the Erdös–Rényi random graph. Obviously, it is also more vulnerable than a regular mesh for which a targeted attack cannot be properly defined (all vertices have the same degree) and thus can be considered to have the same resilience as in the random removal case. The different behavior of the IR graph and the Erdös–Rényi graph can be easily understood in terms of their degree distributions. The scale-free nature of the IR graph makes the long tail of large degree vertices, the hubs, extremely important for keeping the graph connected. Their removal leads right away to network collapse. The Erdös–Rényi graphs, however, have a Poisson degree distribution in which the probability of finding a large hub is exponentially small. Statistically, the vertices' degree is almost constant around the average value, and a targeted removal will soon be equivalent to a random removal of vertices with nearly identical degree. In the latter case a targeted attack, while still performing better than a random removal, does not reach the extreme efficiency achieved in scale-free networks.

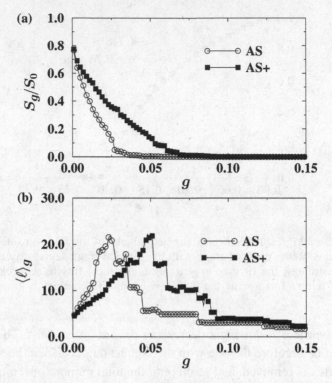

Fig. 6.10 Topological response of the AS and AS+ maps to the targeted removal of the most connected vertices. (a) Relative size of the largest connected cluster in the network with respect to the size of the undamaged network. (b) Average shortest path length of the largest cluster in the damaged network.

The targeted attack scenario is confirmed by the study of the AS and AS+ maps, as shown in Figure 6.10. Also in this case it is observed that the networks almost completely break down at a very small value of g. In particular, the largest cluster in the AS map reaches a value close to zero for the removal threshold $g_c \approx 0.04$. The AS+ map, however, vanishes for $g_c^+ \approx 0.06$. The average shortest path length grows for small g until reaching a maximum value that is, in both cases, close to ten times larger than the value corresponding to the undamaged network. After this maximum, $\langle \ell \rangle_g$ quickly decreases towards zero.[14] The analysis of efficiency, shown in Figure 6.11, confirms this picture. The value of E_g drops very quickly, reaching a value close to zero for a fraction g close to 0.04. Interestingly, the efficiency E_g for the targeted attack seems to decay as the logarithm of g, as shown in the log-linear plot in Figure 6.11. This effect, which does not appear to be present

---

[14] The plateau in the average shortest path length in Figure 6.10(b) is due to the long persistence of a largest cluster of very small size, which finally disappears at larger values of g.

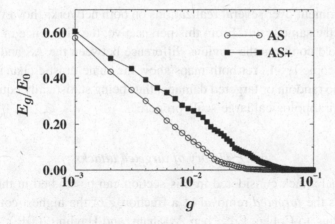

Fig. 6.11 Relative efficiency of the AS and AS+ maps as a function of the fraction of removed vertices in a targeted attack.

in the corresponding plots for S_g and $\langle \ell \rangle_g$, is most probably related to the scale-free nature of the maps.

As a final remark, it is interesting to notice the seemingly different behavior that the AS and AS+ maps show in front of a targeted deletion of vertices. From Figure 6.10 it is apparent that the AS+ map is slightly more robust than the AS map: Its largest cluster disappears for a slightly larger value of g, the peak in $\langle \ell \rangle_g$ being correspondingly shifted to the right. Analogously, the efficiency E_g, Figure 6.11, seems to decrease faster in the AS map than in its enriched counterpart. This difference, which was not present in the case of random damage, is probably due to the fact that, both maps having a very similar number of vertices, AS+ has close to 50% more edges. Therefore, the AS+ map has a larger average degree, which makes it capable of sustaining a slightly larger degree of targeted damage. Moreover, in a targeted attack it is natural to expect differences from map to map, despite their statistical similarities. Indeed, a targeted attack consisting in deleting nodes in a sequence of decreasing degree, in an essentially deterministic procedure,[15] is not a self-averaging process and is thus very sensitive to the fine details of the map under consideration. A random attack, on the contrary, is self-averaging, since it is possible to have a statistical sampling of various damage realizations in the same network for a fixed value of g. For instance, two artificially generated graphs with the same statistical properties could just as well result in slightly different thresholds due to the particular graph realizations considered. Averaging the random

---

[15] One might consider the possibility of averaging a targeted attack experiment with a given value of g, that deletes gN vertices and in which the first surviving vertex has degree k_{max}, over all the possible combinations of vertices with degree k_{max} to be deleted, if there is more than one. The effect of this average, in any case, will be always much smaller than in the case of random removal.

damage experiment over several realizations on both networks, however, will produce exactly the same result. From this perspective, the difference in the targeted attack threshold confirms the obvious difference between the AS and AS+ maps at the microscopic level. Yet both maps show the same drastic change of picture in resilience to random or targeted damage, thus being statistically equivalent with respect to their topological large-scale properties.

6.6.1 Theory of targeted attacks

The intentional attack considered in this section can be studied in the context of percolation as the *targeted* removal of a fraction g of the highest connected vertices. According to Cohen, Erez, ben Avraham, and Havlin (2001a), the removal of the most connected vertices has two parallel effects: (i) it reduces the original cut-off k_c of the degree distribution, and (ii) it deletes all the edges connected to the removed vertices. The first effect can be easily taken into account by considering that, after the targeted deletion, there is a fraction $1 - g$ of surviving elements, such that[16]

$$1 - g = \sum_{k=m}^{k_c(g)} P_0(k), \tag{6.34}$$

where m is the network's minimum degree and $k_c(g)$ is the new cut-off of the degree distribution due to the removal of high degree nodes. Let us now specifically consider the study of random uncorrelated scale-free graphs with degree distribution $P(k) \sim k^{-\gamma}$, which is the case of interest because of the drastic change of resilience properties. In this case, neglecting the contribution of k_c, that is supposed to be very large compared with $k_c(g)$, we find

$$g = \sum_{k=k_c(g)+1}^{\infty} P_0(k), \tag{6.35}$$

that, in the continuous k approximation, can be inverted to yield

$$k_c(g) \simeq m g^{1/(1-\gamma)}. \tag{6.36}$$

That is, the surviving subgraph has a cut-off $k_c(g) < k_c$, that is a decreasing function of g, and diverges for $g \to 0$. In order to take into account the reduction in the number of edges implicit by the deletion of vertices, we consider that, in a random uncorrelated graph, the probability that an edge points to a vertex of degree k

---

[16] Note that in the analytic treatment of targeted attack we do not face the deterministic problem found in numerical experiments in finite networks, since we are dealing with quantities averaged over an ensemble of statistically equivalent random graphs of infinite size.

is $k P_0(k)/\langle k \rangle_0$ (for the undamaged distribution). Therefore the probability that an edge is removed since it is pointing at a removed vertex is

$$r(g) \simeq \frac{1}{\langle k \rangle_0} \sum_{k=k_c(g)+1}^{\infty} k P_0(k). \qquad (6.37)$$

Neglecting again the contribution of k_c, and using the relation (6.36) between g and $k_c(g)$, we find that

$$r(g) \simeq g^{(2-\gamma)/(1-\gamma)}. \qquad (6.38)$$

Cohen *et al.* (2001a) argue that, after a targeted removal of the more connected vertices, the surviving graph is equivalent to a scale-free network with cut-off $k_c(g)$ in which a *random* removal of *edges* with probability $r(g)$ has been performed. Using the results of Section 6.5 (since they are valid for the removal of either vertices or edges), the threshold value g_c is found as the self-consistent solution of the usual relation for the disappearance of the giant component, i.e.

$$r(g_c) = 1 - \frac{1}{\kappa(g_c) - 1}, \quad \text{with} \quad \kappa(g_c) = \frac{\sum_{k=m}^{k_c(g_c)} k^2 P_0(k)}{\sum_{k=m}^{k_c(g_c)} k P_0(k)}, \qquad (6.39)$$

where the cut-off $k_c(g_c)$ is the one corresponding to the threshold g_c. In the continuous k approximation, we can integrate the previous expression and use Eqs (6.36) and (6.38) to obtain

$$g_c^{(2-\gamma)/(1-\gamma)} = 2 + \frac{2-\gamma}{3-\gamma} m \left(g_c^{(3-\gamma)/(1-\gamma)} - 1 \right), \qquad (6.40)$$

that can be solved numerically to obtain the threshold $g_c(\gamma)$ as a function of the degree exponent γ and the minimum degree[17] m.

In Figure 6.12 we plot the threshold g_c as a function of the degree exponent γ for minimum connectivities $m = 1, 2$, and 3. From this figure we can observe that, for any m, g_c reaches a maximum for a value of γ less that 2.5. The absolute value of the maximum is, however, very small. For $m = 1$, for example, the most resilient graph corresponds to a degree exponent $\gamma \approx 2.29$; in order to destroy its giant component it is sufficient to remove 6.2% of the most connected vertices. In the case of the Internet, however, with $\gamma \simeq 2.1$, the deletion of 4.7% of the nodes yields the collapse of the network. This very small number must be contrasted, however, with the value of the cut-off in the surviving subgraph which, from Eq. (6.36), is $k_c(g_c) \sim 16$. This means that, even for this small threshold,

---

[17] This result strictly applies to power-law degree distributions and depends explicitly on γ. For Poisson graphs the final threshold equation should take into account the details of the degree distribution and would lead to threshold values much closer to the random removal case.

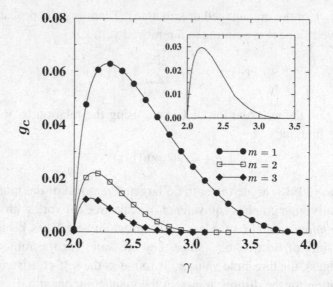

Fig. 6.12 Threshold for the targeted removal of vertices in scale-free graphs with different minimum connectivity m, as a function of the degree exponent γ, computed using the continuous k approximation (Cohen *et al.*, 2001a). The inset shows the values corresponding to the discrete formalism (Dorogovtsev and Mendes, 2001a) for $m = 1$.

great damage has been inflicted to the network's connectivity. It is important to note that the results obtained are defined in a statistical sense. As stressed in the previous sections, targeted attack cannot be averaged on a single graph, since for each graph only a single attack sequence exists. Therefore, the analytically obtained threshold is recovered only by averaging over many network realizations or in the strict infinite size limit. Nevertheless, in large enough networks the single realization threshold will be extremely close to the predicted value.

Finally, the observation that the removal threshold depends considerably on the density of "dangling ends" (vertices with degree one) present in the graph deserves a detailed discussion (Dorogovtsev and Mendes, 2001a). This implies that a more correct description of targeted attacks should take into account the correct form of the degree distribution for small values of k, and in particular its eventual discreteness. For this purpose, Dorogovtsev and Mendes (2001a) consider a *discrete* scale-free degree distribution with minimum connectivity $m = 1$, $P_0(k) = k^{-\gamma}/\zeta(\gamma)$, where $\zeta(x) = \sum_{k=1}^{\infty} k^{-x}$ is the Riemann Zeta function (Abramowitz and Stegun, 1972). Within this discrete formulation, the condition (6.39) can be rewritten as

$$\sum_{k=1}^{k_c(g_c)} k(k-1)k^{-\gamma} = \zeta(\gamma - 1). \qquad (6.41)$$

From this last equation one can numerically obtain the value $k_c(g_c)$, and from it recover the threshold for targeted deletion using Eq. (6.34). The inset in Figure 6.12 shows the threshold g_c thus obtained as a function of the degree exponent. The discrete nature of the degree distribution is reflected in a smaller threshold, which at the maximum point is close to one half the value corresponding to the continuous k approximation. While the effect of the discreteness of the degree distribution appears to be quite noticeable, it is important to keep in mind that, for most real systems such as the Internet, the distribution for small k has not been thoroughly explored. Therefore, and in order to qualitatively account for large k behavior, the continuous approximation qualifies rather well (Cohen, Erez, ben Avraham, and Havlin, 2001b).

6.7 The price of a fail-safe Internet

It is surprising to realize how the original goal for a designed secure computer network has been accomplished is a self-organized Internet, which has not followed any pre-established plan, but has evolved instead driven by local, unpredictable, and sometimes selfish, individual decisions. The outcome of this process is the present Internet, whose scale-free distribution seems to give us for free a stunning robustness to failures. From an economical point of view that is a very welcome property because it allows the use of cheap elements with the certainty that they will not endanger the communication capabilities of the Internet. However, nothing comes for free in the long term, and the price to pay may indeed be very high. The very same scale-free distribution that is responsible for the extreme robustness of the Internet is at the same time its own Achilles's heel, bearing the seed of its possible destruction. A coordinated malicious attack, targeting the most connected vertices, can disrupt with minimal effort the Internet fabric, collapsing it into a myriad of helpless isolated islands, with no capability to communicate among them. This weakness of the Internet points to the need for the urgent design of defense strategies able to protect the Internet from targeted denial-of-service attacks, that could easily cripple the network.

7

Virtual and social networks in the Internet

The Internet and the World Wide Web (also known as WWW or simply the Web) are often considered as synonyms by non-technical users. This confusion stems from the fact that the WWW is at the origin of the explosion in Internet use, providing a very user-friendly interface to access the almost infinite wealth of information available on the Internet. The WWW, though, is a rather different network in the sense that it is just made from a specific software protocol, which allows access to data scattered on the physical Internet. In other words, it is a virtual network which lives only as a sort of software map linking different data objects. Nevertheless, the Web finds a natural representation as a graph and it is a stunning example of an evolving network. New Web pages appear and disappear at an impressive rate, and the link dynamics is even faster. Indeed, the fact that we are dealing with virtual objects makes Web dynamics almost free from the physical constraints acting on the Internet. Any individual or institution can create at will new Web pages with any number of links to other documents, and each page can be pointed at by an unlimited number of other pages.

The Web is not the only virtual network present on the Internet. Users interactions and new media for information sharing can be mapped as well in a graph-like structure. The graph of e-mail acquaintances of Internet users is a well-defined example of social network hosted by the Internet. Similarly, peer-to-peer (P2P) systems, such as the Gnutella file sharing protocol, can be mapped to dynamic graphs with a large number of vertices that join and leave the network at a very high rate. Also in this case, the network is a reflection of a certain particular social community on the Internet.

Virtual networks have became the information transfer media for hundreds of millions of users and, similarly to the physical Internet, have grown to become enormous and complex systems as the result of a self-organized growing process. It is therefore not surprising that we find in virtual networks a vast array of emergent phenomena and topological properties that can be expressed as mathematical laws governing graph structure and evolution. These laws are the outcome of the interactions among the many individuals forming the various communities and can provide insights on many phenomena such as WWW usage and traffic patterns, or spreading of computer viruses by the e-mail.

In this chapter we want to provide an overview of the main virtual networks hosted by the physical Internet. We shall report on the experiments devoted to the measurements of their structure and the results concerning their large-scale topology. The correspondences with the underlying Internet's physical topology as well as the relevant differences in graph structure will be analyzed in order to pinpoint the basic ingredients that must be included in the modeling of virtual and social networks.

7.1 The World Wide Web

7.1.1 What is the Web

The Web, born at the Counseil Europeen pour la Recherche Nucleaire (CERN), is an evident example of the unexpected practical benefits of basic research. The CERN is a very large research facility with thousands of researchers and hundreds of computer systems. In order to help with the sharing of data among researchers in the High Energy Physics Department, Tim Berners-Lee completed in 1989 a research proposal for a system aimed at sharing and navigating information across different networks. Robert Cailliau, who independently presented a project on a hypertext system at CERN, soon joined Berners-Lee in his effort to get the Web off the ground, and in 1990 the first Web browser and server were communicating.

In very simple terms, the Web is an Internet-based computer network that allows the navigation and retrieval of data scattered on the global Internet. It is based on a *client–server* architecture. In this architecture the client relies on a program, the Web browser, that connects to a remote machine, the Web server, where the data are stored. The Web server is the computer that handles all the communication with individual users and were data are stored in the form of Web pages. The structure of Web pages is defined with a common HyperText Markup Language (HTML) so it is easily and quickly communicated across the Internet. Hypertexts point to data objects, be it text or figures, by *hyperlinks* which specify the logical address of the Web page containing the related information. The address space is specified by a Uniform Resource Locator (URL) that identifies Web servers on the Internet and the particular page in each Web server in a hierarchical way very similar to the Domain Name System (see Chapter 2). The communication protocol devised for the WWW is named HyperText Transfer Protocol (HTTP) and allows the downloading and formatting of Web pages by sending a specific request to the Web server. By its nature, the hypertext structure defines a network of pages connected by hyperlinks, readily providing a graph representation whose vertices are HTML pages and edges are the hyperlinks pointing from one page to another.

Two main ingredients contribute to the incredible success of the WWW. First it is a very user-friendly and open system, that can be used by anyone familiar with a computer window and the mouse. In practice, the user just has to follow links which are as natural and intuitive to use as pressing a button. A second important feature is that the program code was in the public domain since its early development stages, so that anyone in the cyber community could use and improve it. These ingredients have made the Web so successful that its rapid and unregulated growth has led to a huge and complex network for which it is difficult to even guess the total number of Web pages.[1] It is not just the structure of the Web that has developed to become very complex and intricate, but also the various dynamics taking place on it, such as navigating patterns, community structures, congestions, and other social interaction phenomena driven by the Web users (Huberman and Lukose, 1997; Huberman, Pirolli, Pitkow and Lukose, 1998). All these factors have triggered the interest of researchers to what has been defined by Huberman (2001) as an ecology of information, and have led to the development of methods for the gathering of data on the graph structure of the Web and the dynamical patterns related to its use.

7.1.2 Structure and topology of the Web graph

All the experiments aimed at studying the graph structure of the WWW are based on Web crawlers that explore the Web connectivity by following the links found on each page. In practice, Web crawlers are special programs that, starting from a source page, detect and store all the links emanating from it and follow them to build up the set of pages reachable from the starting one. This process is then repeated for all pages retrieved, obtaining a second layer of vertices and so on, iterating for as many possible layers as allowed by the available storing capacity and CPU time. From the collected data it is then possible to reconstruct a graph representation of the Web by identifying vertices with Web pages and edges with the connecting hyperlinks.

The Web graph has a basic difference with respect to the graphs analyzed so far in studying the physical Internet. For each Web page we know the number of outgoing hyperlinks, but in principle we know nothing about the incoming hyperlinks from other pages. That is, we can follow outgoing hyperlinks to reach

---

[1] Lawrence and Giles (1999) estimated in February 1999 a lower bound of 8×10^8 documents in the publicly indexable WWW, scattered on about 3 million Web sites.

pointed pages, but we cannot navigate on the way back the incoming hyperlinks. The Web graph is thus a *directed* graph, in which edges connect ordered pairs of vertices (see Appendix A1). This implies that, given a pair of vertices (i, j), we must specify if the connecting edge goes from i to j or vice-versa. The directness has several noticeable implications on graph characterization. First of all, we can introduce the out-degree k_{out} and the in-degree k_{in} of each vertex, defining the number of outgoing and incoming directed edges, respectively. In addition, the connectivity properties depend on the direction of the edges. A path connecting two vertices i and j exists if there is a sequence of directed edges going from i to j through an intermediate number of vertices. However, a path from i to j does not necessarily imply the presence of a path from j to i. This fact complicates reasonably enough the structure of the connected components in directed graphs, as shown in Appendix A1. Moreover, the directness affects the definition of the average shortest path length of the graph $\langle \ell \rangle$, since we have to restrict the average over pairs of vertices for which a directed path exists.[2] For this reason, in directed graphs it is sometime preferable to consider the *diameter* d_G of the graph, defined as the largest of the shortest directed paths.

One of the first attempts to characterize the topological properties of the Web graph has been provided by Albert *et al.* (1999) by analyzing data collected by a specifically devised Web crawler, mapping 325,729 Web pages and 1,469,680 hyperlinks in the University of Notre Dame domain. This analysis highlighted the small-world property of the Web, that despite the large number of pages considered showed an average shortest path $\langle \ell \rangle \simeq 11$. This striking evidence for the small-world character of the WWW has been confirmed by the largest data sample analyzed (see Broder, Kumar, Maghoul, Raghavan, Rajagopalan, Stata, Tomkins, and Wiener, 2000). In this work, the authors used databases from a Web crawl performed from several starting points at AltaVista in May/October 1999.[3] Because of the multiple starting points and a full account of the directed nature of the edges, this study identified the full complex hierarchy of connected components allowed by the directed nature of the Web graph. In Figure 7.1 the full structure of the largest connected component of the Web graph is reported. In particular, the giant strongly connected component (GSCC), where a directed path exists between any pair of pages, is formed by 56 million pages. Connected to this core the IN and OUT components, also referred to as the giant in- and out-components, are found. These two sets are formed by pages linked by a directed path that enters into or exits from the GSCC, respectively, and amount to 44 million pages each. Along

---

[2] Pairs not connected with a directed path are considered to have infinite distance, and even a single pair of this kind would cause the average shortest path length to diverge.

[3] The largest of these databases contains 271 million pages and 2,130 million hyperlinks.

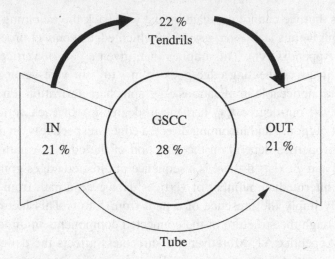

Fig. 7.1   Sketch of Web connectivity. The giant weakly connected component, composed of the pages connected without regard to the directed nature of edges, is partitioned into several sets. From any vertex in the IN set it is possible, passing through vertices in the GSCC, to reach any vertex in the OUT set. Tendrils contain vertices that are reachable from portions of the IN set, or can reach portions of the OUT set. Finally, tubes represent a direct passage from some vertex in IN to some vertex in OUT without passing through the GSCC. The remaining 5% of vertices belong to smaller disconnected components. The shape of the figure has led to the definition of *bow-tie* connectivity structure. Figure adapted from Broder *et al.* (2000).

with these larger sets, smaller disconnected components, as well as tendrils and other structures, have been identified in the graph.[4]

In these datasets, the average shortest path length between vertices connected by a directed path is slightly over 16, in fair agreement with the small-world behavior expected for graphs of size $N \simeq 10^8$. It must be noticed, however, that the significance of this quantity is diminished by the fact that, for the majority of pairs of pages, there is no directed connecting path. The small world property, however, is clearly recovered in the directed diameter of the GSCC that settles into the very small value of 28. Finally, it is interesting to observe that the small-world properties persist at a different level of granularity. For instance, Adamic (1999) has studied the WWW at the level of Web sites, thus coarse graining many pages into just one vertex of the representative graph. Also in this case it is found that the average distance between any two Web sites is just about four clicks. Interestingly, it is at the level of Web pages that we can find a first signature of the presence of scale-free features in the Web. That is, the number of Web pages contained by Web sites is highly variable, with a few sites containing even thousands of pages

---

[4] See Appendix A1 for their precise definition.

while a majority of them contain just a few tenths. Not surprisingly, the probability distribution that a site has a certain number of pages is a heavy tailed distribution well approximated by power-law behavior (Adamic and Huberman, 2001).

As for the physical Internet, relevant topological information and further evidences for the scale-free character of the Web can be gathered by studying the in-degree and out-degree probability distributions. The behavior of the probability distribution $P(k_{in})$ that any given Web page is pointed by k_{in} hyperlinks shows a remarkable agreement in all studies, regardless of the sample size considered (Kumar, Raghavan, Rajagopalan and Tomkins, 1999; Albert *et al.*, 1999; Broder *et al.*, 2000; Laura, Leonardi, Millózzi, Meyer, and Sibeyn, 2003). The distribution exhibits a heavy tailed form, well approximated by power-law behavior $P(k_{in}) \sim k_{in}^{-\gamma_{in}}$, with $\gamma_{in} \simeq 2.1$ (see Figure 7.2). The out-degree distribution also appears to be heavy tailed. However, attempts to fit the distribution with strict power-law behavior provide exponent values ranging from 2.4 to 2.8, depending

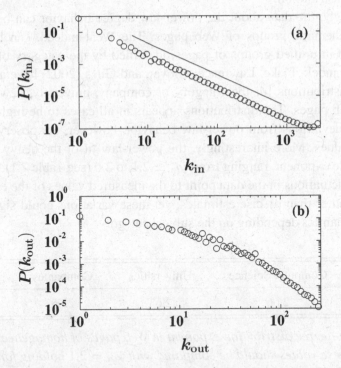

Fig. 7.2 Degree distributions of (a) the in-degree and (b) the out-degree of WWW pages. The double logarithmic plot evidentiates the heavy tailed properties of both distributions. For the in-degree distribution a very good power-law fit $P(k_{in}) \sim k_{in}^{-\gamma_{in}}$ with $\gamma_{in} = 2.1 \pm 0.1$ is obtained. The degree data are from Laura *et al.* (2003), obtained by analyzing a Web graph collected by Gary Wesley in December 2000 at the WebBASE project at Stanford University (http://www-diglib.stanford.edu/~testbed/doc2/WebBase/).

on the range of k_{out} and the Web sample considered.[5] In particular, in the largest data sets analyzed (Broder *et al.*, 2000; Laura *et al.*, 2003), the out-degree distribution appears to have an evident bending deviating from a pure power-law behavior. As discussed in Section 4.4, this bending could be the signature of an exponential cut-off of the distribution. The origin of the cut-off can be explained by the different nature of the in-degree and out-degree evolution. The in-degree of a vertex is the sum of all the hyperlinks incoming from all the Web pages in the WWW. In principle, thus, there is no limit to the number of incoming hyperlinks, that is determined only by the popularity of the Web page itself. On the contrary, the out-degree is determined by the number of hyperlinks present in the page, which are controlled by Web administrators. For evident reasons (clarity, handling, data storage) it is very unlikely to find an excessively large number of hyperlinks in a given page. This represents a sort of finite capacity (Amaral *et al.*, 2000) for the formation of outgoing hyperlinks that might naturally lead to a finite cut-off in the out-degree distribution.

Interestingly, the data about the power-law degree behavior can be refined by looking at thematic groups of Web pages. The Web, indeed, can be naturally decomposed in unified groups of pages, identified by the category of the pages' contents. Pennock, Flake, Lawrence, Glover, and Giles (2002) have analyzed the in-degree distributions for the category of company, university, newspaper, and scientist Web pages. The distributions appears in all cases to be highly variable. Systematic deviations from power-law behavior, however, are observed at small in-degree values. More interestingly, the power-law fit of the heavy tail gives a quite variable exponent, ranging from $\gamma_{in} \simeq 2.1$ to 2.6 (see Table 7.1). Despite the statistical fluctuations in the data point to the measured values of the exponents as indicative rather than precise estimates, yet these variations could signal slightly different dynamics depending on the subject category.

Category	Computer Science	Universities	Companies	Newspapers
γ_{in}	2.66	2.63	2.05	2.05

Table 7.1 *In-degree distribution exponent in Web pages of homogeneous category. These values should be compared with $\gamma_{in} = 2.1$ holding for the whole Web statistics (Data from Pennock* et al., *2002)*

[5] Strikingly, the power-law behavior of the in-degree and out-degree distributions is recovered also at the coarser granularity of Web sites. Also in this case the probability that any site has a given aggregated number of hyperlinks pointing to or departing from it is power-law distributed over almost five orders of magnitude (Adamic and Huberman, 2001).

The obvious existence of thematic groups in the WWW can be extended by considering the large number of communities hosted by the Web. These are groups of people sharing the same interests, which represent the sociological dimension of the Web. This characterization is obviously of great importance in the development of Web navigation tools and search engines. The vast activity on the "taxonomy" of Web communities is a field of research *per se*, that struggles with the measuring of the the semantic content of the pages as well as the link topology. While an extensive account of these studies is beyond the scope of this book, it is worth mentioning the attempts to find a signature of various cyber communities that have led to several analyses focusing on subgraphs of the WWW graph (Gibson, Kleinberg, and Raghavan, 1998; Kumar *et al.*, 1999; Flake, Lawrence, and Giles, 2000; Kleinberg and Lawrence, 2001; Adamic and Adar, 2001; Eckmann and Moses, 2002). In general, communities are identified by an unusually high density of edges among small subgraphs (see Figure 7.3). Given the directed nature of the WWW, a first mathematical way to account for these communities is to look at the number of bipartite cliques present in the graph (Kumar *et al.*, 1999; Dill, Kumar, McCurley, Rajagopalan, Sivakumar, and Tomkins, 2001; Laura, Leonardi, Caldarelli, and De Los Rios, 2002). A bipartite clique $K_{n,m}$ identifies a group of n vertices, all of which have a direct edge to the same m vertices. Naïvely, we can think of the set as a group of "fans" with the same interests and thus pointing in their Web pages to the same set of relevant Web pages of their "idols." Another way to detect communities is to look for subgraphs where vertices are highly interconnected among themselves and poorly connected with vertices outside the subgraph. In this way, different communities can be traced back with respect to

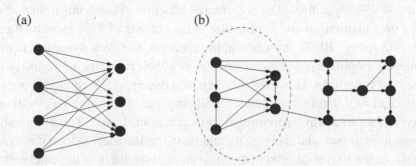

(a)   (b)

Fig. 7.3 Communities of pages on the same topic can be identified by highly interconnected subgraphs. (a) A clique $K_{4,3}$ in which four pages of *fans* point to the same set of pages, the *idols*. (b) A community of vertices (within the dashed oval) weakly connected to the rest of the network. A community is also detectable as a set of pages in which each vertex has a higher density of edges within the set than with the rest of the network. Figure adapted from Kleinberg and Lawrence (2001).

varying levels of cohesiveness. In general, the Web graph presents a high number of bipartite cliques and interconnected subgraphs, all identified by an unusually high density of edges. It easy to realize that such local properties are likely to correspond to a high clustering coefficient for the subgraphs, and the global statistical abundance of communities can be obtained by analyzing the average clustering coefficient of the Web graph. Measurements of the average undirected clustering coefficient[6] confirm the clustered nature of the Web, providing experimental values about four to five orders of magnitude larger than those expected in random graphs of comparable size (Laura *et al.*, 2002). The Web is thus no exception to the small-world-yet-clustered nature of many social and technological networks.

In summary, the WWW dynamical evolution generates small-world properties, heavy tailed distributions, and a high level of clustering along with a rich connectivity structure due to the directness of the graph. As for the physical Internet, the Web too cannot find a satisfactorily representation in terms of standard static models. Also in this case a correct modeling calls for the identification of the basic organizing dynamical principles spontaneously leading to the wide range of emergent properties observed in real data analysis.

7.2 Modeling the Web

The WWW and the Internet graphs exhibit qualitatively similar topological features. This suggests that, despite their different nature, both networks are possibly following some common organizing principles. Indeed, both graphs belong to the more general class of growing scale-free networks and many of the models aimed at describing these networks have been originally devised having in mind the WWW dynamics. The preferential attachment mechanism, in particular, finds its very inspiration and application in the context of Web growth (Barabási, Albert, and Jeong, 2000). At a first approximation, the Web dynamics is eventually ruled by popularity. The more popular is a Web page, the more likely it will receive new hyperlinks. However, an obvious measure of a Web page's popularity is the number of pointing hyperlinks, readily implying that a Web page will receive new hyperlinks in a way proportional to its actual in-degree. The preferential attachment mechanism, and its implementation in the Barabási–Albert construction, thus indicate the universal growth mechanism as at the basis of the emergent properties of both the Internet and the WWW. However, the Barabási–Albert algorithm is just the simplest implementation of the preferential attachment mechanism and it is not intended to be a realistic model of any real-world network. Rather, it is a zero-order conceptual model which can be used as the paradigm for much more

---

[6] The directionality of edges is not considered.

articulate models, taking into account the various particular processes taking place in the network under consideration. In the following we will provide a review of models that consider explicitly some of the specific characteristics of the WWW dynamics.

7.2.1 Preferential attachment models

As we have seen for the Internet, a wide range of growing models have been formulated, considering extra features such as rewiring, additional edges, and fitness heterogeneity (see Chapter 5). All these additional ingredients are also likely to be relevant in the WWW evolution. Realistic models for this network, however, cannot overlook the directed nature of this graph. This has led to the formulation of several growing network models in which the directionality of edges is explicitly considered in the evolution dynamics.

A first example in this direction is the directed model proposed by Dorogovtsev *et al.* (2000) (see Section 5.6). Essentially this model considers a Barabási–Albert growth rule with directed edges. At each time step a new vertex is introduced and emanates m directed edges to already existing vertices following a preferential attachment with probability $\Pi[k_{in}] \sim A + k_{in}$. The variable A is a constant representing the initial attractiveness of each vertex, and allows new vertices with no incoming edges to participate in the dynamics. The growth rate equation for the average in-degree at time t of a vertex appearing at time s thus reads as

$$\frac{\partial k_{in,s}(t)}{\partial t} = m \frac{A + k_{in,s}(t)}{(A + m)t}, \tag{7.1}$$

where the denominator is the correct normalization factor due the t vertices present in the network. The solution to this model is analogous to those shown in Section 5.6 with the simple rescaling $2m \to m$ in the normalization factor, since directed edges count just once in the in-degree. Taking into account the boundary condition $k_{in,s}(s) = 0$ (all vertices are added with a zero in-degree), we obtain the in-degree distribution $P(k_{in}) \sim (A + k_{in})^{-2-A/m}$. The introduction of constant popularity has thus the advantage of introducing the possibility of tuning the degree exponent to any value in the range $]2, \infty]$. With this definition, however, the model yields an out-degree distribution that is a delta function at the value m, i.e. $P(k_{out}) = \delta(k_{out} - m)$. A model with a very similar structure, the addition of a uniform component, has been also proposed by Pennock *et al.* (2002) to fit the degree distribution variability found in uniform topic Web pages.

Tadic (2001) and Krapivsky, Rodgers, and Redner (2001) focus on another distinctive feature of the WWW by developing directed models in which a high level of rewiring is considered. The insight at the basis of these models is that edges between pairs of vertices change on a very short time scale, comparable with that of the new vertices' appearance. The proposed models contain a few differences in the precise formulation of the dynamics which however do not affect the final topology of the obtained graph. Tadic's (2001) version of the model incorporates a growth mechanism in which at each time step a new vertex is added along with M new edges. A fraction αM of these edges emanates from the new vertex, while the rest is established between already existing vertices. The parameter α thus measures the balance between new edges due to the enlargement of the network and those due to updating and discovery of new interesting pages. Since the network is directed, in placing the new edges among existing vertices, the emanating edge has to be specified. This is made stochastically by selecting the emanating vertex i with a probability p proportional to the vertex out-degree $k_{out,i}$; i.e., $p(i) \sim k_{out,i}$. This amounts to assuming that the updating of edges is occurring more frequently on vertices with a large number of outgoing hyperlinks. These vertices are supposed to be more active and thus more frequently selected. New edges from both new and old vertices point to vertices chosen by following the usual preferential attachment rule in which the popularity is measured as the number of incoming edges; i.e. $\Pi[k_{in}] \sim k_{in}$. The model can be inspected both analytically and numerically and yields power-law degree distributions $P(k_{in}) \sim k_{in}^{-\gamma_{in}(\alpha)}$ and $P(k_{out}) \sim k_{out}^{-\gamma_{out}(\alpha)}$, where the degree exponents are a function of the parameter α expressing the rate of edges due to new vertices. The parameter α allows the tuning of the exponents and can be considered as an external input to be obtained by empirical measurements.

The previous models consider that the total number of edges present in the network is a monotonous growing function of time. This is true on average due to the net balance between the appearance and updating of new hyperlinks and their disappearance when they become obsolete. However, if a realistic Web dynamics is the final aim, the latter processes should be explicitly considered. This is indeed the case of the directed version of the Internet model by Goh *et al.* (2002) reported in Section 5.8.1. Inspired by the multiplicative noise scenario put forward by Huberman and Adamic (1999), the model assumes that the number of vertices N follows a steady multiplicative growth of the form $\frac{\partial N(t)}{\partial t} = pN(t)$, that in a discrete formulation implies that at the time step t, $pN(t-1)$ new vertices are introduced in the network (Kahng, Park and Jeong, 2002). Each one of these new vertices then emanates m edges pointing to old vertices chosen following the preferential attachment rule for the in-degree, i.e. $\Pi[k_{in}] \sim k_{in}$. Since vertices have the possibility of receiving new edges only because of their popularity, measured by $k_{in,i}$, new vertices are assigned an entry level of popularity equal to 1. The updating of hyperlinks is then represented by the time evolution of the out-degree, which follows for each vertex the dynamics

$$\frac{\partial k_{out,i}(t)}{\partial t} = k_{out,i}(t)\xi_i(t). \tag{7.2}$$

Here $\xi_i(t)$ is a stochastic variable with average $\langle \xi_i(t) \rangle = g$ and variance $\langle \xi_i(t)\xi_{i'}(t') \rangle^2 - g^2 = \sigma^2 \delta_{t,t'}\delta_{i,i'}$. The parameter $g > 0$ represents the average growth rate of outgoing hyperlinks and σ is the fluctuation level of this process. This implies that, while on average the number $k_{out,i}$ is increasing, negative fluctuations may lead to the deletion of edges. In particular, if $k_{out,i}\xi_i(t) > 0$, the same number of new outgoing edges will be established with randomly chosen vertices following the preferential attachment rule $\Pi[k_{in,i}] \sim k_{in,i}$. If $k_{out,i}\xi_i(t) < 0$, the corresponding number of vertices will be simply deleted. The model can be analytically studied, yielding a power-law behavior for both the out- and in-degree distributions. While the out-degree exponent depends in general on the full set of parameters p, g, and σ, the in-degree exponent can assume only two values, namely $\gamma_{in} = 2$ if $p \geq g$ and $\gamma_{in} = 1$ if $p < g$. This result appears particularly interesting since it relates the topology of the Web to the relative growth rate of vertices and edges. In particular, the experimental results appear to place the WWW in the case $p \geq g$, in which the number of Web pages is increasing faster than the average number of outgoing edges. This seems perfectly plausible because, while the number of Web pages is increasing at a very quick pace that precludes even their numerability (Lawrence and Giles, 1999) the number of hyperlinks that can be added to each page is constrained by handling and visual representation capabilities and our limited knowledge of the Web.

A common characteristic of the WWW models analyzed so far is that the heterogeneity of vertices is not considered. This implies that they are differentiated only by the arrival time. As we discussed in Section 5.8.2, this makes older vertices always the most connected ones. Implicitly, we are not considering that a new Web page with very interesting content may rapidly become a very successful one and outmatch older competitors. On the contrary, we know that this kind of event often occurs, and Adamic (2001) has provided a quantitative measure of this effect by showing that in-degree and age are rather uncorrelated variables in WWW pages. The natural way to deal with this fact in Web models is with the introduction of a fitness parameter that models the heterogeneous attractiveness of Web pages (Bianconi and Barabási, 2001). The original fitness model was for undirected networks, and a promising path for WWW modeling is represented by the inclusion of directness elements in its formulation.

Finally, we want to mention an interesting model which combines the preferential attachment paradigm with the textual content affinity of Web pages (Menczer, 2002). Indeed, it is easy to see that popularity is not the only force that drives the establishment of hyperlinks. Despite the fact that the *Internet Movie database* Web site is extremely popular, it is very unlikely that a Web page specializing in *gardening* will point to it. Therefore, the in-degree popularity has a limited significance out of thematic areas. In particular, Menczer (2002) has empirically measured the probability that two given pages i and j are connected as a function of their *lexical distance*[7] $\rho(i, j)$. The analysis of real Web samples indicate that below a certain threshold value ρ^*, the probability that two pages are connected is independent on the lexical distance, while above the threshold this probability decays as a power law. Menczer (2002) has cast this empirical evidence in the following definition of a growing network model: at each time step a new vertex (Web page) i joins the network. To each new vertex it is assigned a lexical distance $\rho(i, j)$ with all existing vertices j. These values are randomly drawn from a probability distribution $p(\rho)$ empirically derived from real data. The new vertex then emanates m new edges pointing to already existing pages selected with probability

$$\Pi[k_{in,j}] = \begin{cases} k_{in,j}/mt & \text{if } \rho(i, j) < \rho^* \\ c\rho^{-\alpha}(i, j) & \text{otherwise.} \end{cases} \tag{7.3}$$

Here, α, c and ρ^*, analogously to the lexical distance probability distribution $p(\rho)$, are external parameters to be consistent with the empirical measurements and normalization factors. The hyperlink addition is thus based on the pages' popularity only for pages with similar content, and assumes a decreasing power-law probability for pages with very different content. The model is quite susceptible to the external parameters introduced, but it appears to reproduce quite well the topology of the data sets from which these values are consistently obtained.[8] Though, the most important feature of the model consists in the introduction of a locality mechanism in the growth evolution. The model does not assume that the author of a given Web page (the vertex) has a global knowledge of the whole network popularity (the in-degree of all other vertices). On the contrary, it assumes the knowledge of popularity only for pages with similar content. This is a plausible assumption as content-related pages are known to the author or easily discovered by using search engines. We shall see in the next section how limited knowledge assumptions might be used to effectively recover a preferential attachment mechanism without requiring its explicit implementation in the growth process.

---

[7] The lexical distance is measured in terms of the textual content similarity of the pages with metrics traditionally used in information retrieval (Menczer, 2002).

[8] Menczer (2002) reports numerical simulations with parameter values obtained by data collected on 150, 134 Web pages extracted from the Open Directory Project snapshot of February 14, 2002, http://dmoz.org.

7.2.2 Alternative mechanisms for preferential attachment

A different class of models relies on plausible dynamical mechanisms for Web evolution that at first sight seem to depart from the preferential attachment mechanism. However, as we shall see in the following, they do contain this mechanism in disguise, and probably represent a basic understanding of its very microscopic origin in the WWW.

The copying mechanism is one of the alternative processes considered in the WWW dynamics (Kleinberg, Kumar, Raghavan, Rajagopalan, and Tomkins, 1999; Kumar, Raghavan, Rajagopalan, Sivakumar, Tomkins, and Upfal, 2000). The inspiring consideration for this mechanism is that new pages dedicated to a certain thematic area copy hyperlinks from already existing pages with similar content. This is because Web page authors find out about Web pages of related content and follow their hyperlinks, many of them pointing to other related Web pages. Naturally, a certain fraction of these hyperlinks will be copied also in the new page. This has been translated in a growing model in which at each time step a new vertex (Web page) is added to the network[9] and a corresponding *prototype* vertex is selected at random among those already existing. Each new vertex emits $m \geq 1$ new outgoing edges initially chosen to point to the vertices pointed to by the prototype vertex. At this stage a *copy factor* α (constant for all new vertices) is introduced. With probability $1 - \alpha$ each edge is retained as it is; with probability α it is rewired on a randomly chosen vertex of the network. The copy factor introduces the possibility that not all edges are just copied from the prototype vertex, since the Web page author might find other interesting pages in the network by a random exploration. A pictorial illustration of the copying model dynamics is provided in Figure 7.4.

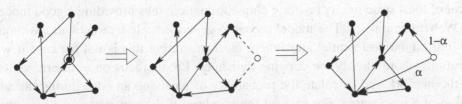

Fig. 7.4   Illustration of the rules of the copying model. A prototype vertex (black dot surrounded by a circle) is selected and a new vertex (hollow dot) is created with virtual edges pointing to the neighbors of the prototype. With probability $1 - \alpha$ the virtual edges are kept; with probability α they are rewired to a randomly chosen vertex.

---

[9] Along with this linearly increasing model, Kumar *et al.* (2000) consider also a model in which there is an exponential increasing of the number of vertices in the network; i.e. at each time step a number of vertices corresponding to a fraction of the entire network is introduced. The resulting network topology is qualitatively similar to the linear growth case described here.

The copying model has been studied analytically, and Kumar *et al.* (2000) have provided rigorous results concerning the in-degree distribution and the number of cliques present in the network. While the out-degree distribution is by construction $P(k_{out}) = \delta(k_{out} - m)$, the in-degree distribution is power-law distributed according to $P(k_{in}) \sim k_{in}^{-(2-\alpha)/(1-\alpha)}$. The copy factor is thus the tuning parameter for the degree exponent of the model. It also controls the number of cliques formed by the network. By definition, the copying model favors the formation of cliques. Indeed, duplicated vertices naturally lead to bipartite cliques pointing to the same set of vertices. This is evident for all cliques $K_{i,j}$ with $j \leq m$, whose formation and enlargement is regulated by the probability that a vertex duplicates the j edges of a "fan" leading to the "idols," i.e. $(1 - \alpha)^j$. In particular, it has been shown that the number of cliques $K_{i,m}$ increases as a power law with the network size. This is very different from what happens in random graph models where the number of cliques is very small and constant with the graph size.

Given its richness and fundamental character, the copying model has also stimulated the development of models enriched by other features, aimed at a more detailed representation of the WWW graph. Among the various generalizations, a very interesting one consists in the layer model proposed by Laura *et al.* (2002). In this model, each new vertex is assigned to one or more thematic layer and the copying mechanism takes place only with vertices belonging to the same thematic layers. The definition of thematic layers allows a richer community structure and the introduction of a content characterization in the formation of the Web graph.

The copying mechanism generates power-law in-degree distributions and naturally leads also to the small-world property in view of the general results of Section 5.6 for random scale-free models. Additionally it also generates a noticeable level of local structure by favoring clique formation, thus providing a good model of Web communities. The model seems to generate scale-free networks without relying on the preferential attachment mechanism, but this is not the case if we scrutinize more closely the copying dynamics. Let us focus on a generic vertex of the network and calculate the probability of receiving an edge during the addition of a new vertex. For each of the m edges of the new vertex we have that with probability α a random vertex in the network is chosen. Thus any vertex has a probability α/N to receive an edge, where N is the size of the network. With probability $1 - \alpha$, however, the vertex, which is pointed to by one of the edges of the prototype vertex, is selected. The probability that any given vertex is pointed to by this edge is given by the ratio between the number of incoming edges of that vertex, and the total number of edges, i.e. $k_{in,s}/mt$. This second process increases the probability of high degree vertices to receive new incoming edges and in the limit of large network sizes we have that the mean-field evolution for the copying model can be written in the usual growth rate equation as (Section 5.4)

$$\frac{\partial k_{in,s}(t)}{\partial t} = m \left[\frac{\alpha}{t} + (1 - \alpha) \frac{k_{in,s}(t)}{mt} \right], \tag{7.4}$$

where it is considered that $N \simeq t$ for large linearly growing networks.

The solution of the above equation with the boundary condition $k_{in,s}(s) = 0$ yields

$$k_{in,s}(t) = \frac{\alpha m}{1 - \alpha} \left[\left(\frac{t}{s} \right)^{1-\alpha} - 1 \right]. \tag{7.5}$$

From this equation, the in-degree distribution can be easily found by applying the general method shown in Section 5.4, namely

$$P(k_{in}) \sim (k_0 + k_{in})^{-\frac{2-\alpha}{1-\alpha}}, \tag{7.6}$$

where $k_0 = \alpha m/(1 - \alpha)$ is an offset constant, thus recovering for large k_{in} values the rigorous result of Kumar *et al.* (2000). Therefore, through its local dynamical rules, the copying model defines an effective preferential attachment growth dynamics. This is a striking result, since the model is defined on the very simple assumption of selecting a prototype vertex, without any knowledge of the popularity or the degree importance of the vertex. The copying model thus offers a microscopic explanation for the preferential attachment mechanism that was just used as an empirical law in other models.

A Web growth model based on a different local exploration mechanism, though with a mathematical structure similar to the copying model, has been put forward by Vázquez (2001). In this model the basic consideration for the drawing of hyperlinks by the Web page author is the random exploration of the existing network. It is therefore supposed that each time a Web page is created it has a hyperlink to a content-related page. Then the author explores the Web starting with this page and following the outgoing hyperlinks. With a certain probability some of these pages will be content related to the newly created Web page and the author will draw a hyperlink. Each time a new interesting page is found, an exploration starts from that page and the process is recursively repeated. Only when no interesting pages are found (no hyperlinks are drawn) does the process stop. This exploration mechanism is translated into a discrete random exploration model defined as follows (see Figure 7.5). A new vertex is added to the network and an edge is pointed at a randomly chosen vertex among the already existing ones. From the randomly chosen vertex the exploration of one or more of its neighbors is performed. With probability α the new vertex points a directed edge also to the explored neighbors. The process is then repeated from the explored vertices until no new edges are

(a)   (b)

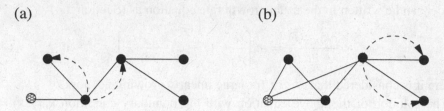

Fig. 7.5   Illustration of the exploration mechanism. (a) The new vertex (shaded circle) draws an edge to a randomly selected vertex of the network. From the selected vertex an exploration process of the neighbors (dashed arrows) is performed. (b) With probability α the new vertex establishes edges with the explored vertices and from them starts a new exploration process. The process stops when no further edges are added to the new vertex.

created. In this case a new vertex is created and the process starts again. The exploration process can proceed in parallel on all the neighbors of the visited vertices or just on a subset of them (Vázquez, 2001, 2002). This model produces a power-law in-degree distribution with an exponent depending on α. In addition, in the case where a complete exploration of neighbors is performed, the distribution presents a transition from a power-law behavior to an exponential one at a threshold value of the probability α. Similarly to the copying model, highly connected vertices have a larger probability to be visited during the exploration process. This can be intuitively understood by noticing that a vertex has a probability to be visited from an exploration that is proportional to the number k_{in} of incoming edges. Also in this case, a local dynamical rule provides a microscopic origin for the preferential attachment mechanism at the global level.

The exploration and the copying models open the path to the inclusion of processes inspired by the social behavior of the agents contributing to Web growth in the large-scale simulation of its evolution. Network modeling can thus find inspiration and cross-fertilization in the noticeable effort that has been carried out to model various dynamical processes occurring on the Web. We shall see in the following chapters problems related to searching and data retrieval, but also usage patterns and download time distributions have been analyzed. An introduction to these arguments can be found in the book by Huberman (2001), which provides a general perspective in the area of multi-agent modeling of WWW phenomena.

7.3 The e-mail network

Each computer in the Internet may use one of the many e-mail programs that allow to write texts and send them to other Internet users. In particular, many Internet hosts run e-mail servers that manage the addresses in their respective domains. Each user address is expressed in the form "name@domain," that is unique at that domain. Therefore, any e-mail server can use the Internet's Domain Name System to retrieve the IP address of any other e-mail server, connect to that server, and transfer e-mail intended to reach recipients at that domain using standard communication protocols. On its turn, the recipient e-mail server will deliver the e-mail to the specific user address.

E-mail is one of the oldest Internet tools, and it has developed a rich range of advanced features. It allows users to exchange long documents as attachments and has one-to-many communication capabilities. These features, along with the real-time velocity of transmission, are at the origin of the success of e-mail, and explain its current status as the most commonly used application on the Internet. Without doubt e-mail has changed the way people communicate and the study of e-mail exchange can therefore provide a great deal of information on how people interact, and reveal many features of the Internet's social structure. This social structure assumes a particular relevance in the study of community organization and the emergence of informal social networks (Guimera, Danon, Diaz-Guilera, Girault, and Arenas, 2002) and has major implications in the spread of computer viruses and worms as we shall see in Chapter 9.

A first method in the study of the e-mail social structure is the characterization of the network induced by e-mail exchange. In this network each vertex corresponds to an e-mail address (user) and edges represent the fact that e-mails are exchanged between two addresses. As usual, this very intuitive mapping finds several complications at the stage of real data collection. In order to have a record of all the interactions among addresses within a given domain we can make use of the local e-mail server. Every time an e-mail is sent or received, a record of the transaction is routinely registered in the *log file* of the e-mail server. By looking at this file it is therefore possible to construct the connectivity of all addresses within the domain. These vertices will also have edges with vertices representing addresses out of the domain for which, however, it is possible to have only limited information. In particular, vertices corresponding to addresses outside the domain will have by definition a degree that is typically underestimated, since many edges representing e-mail exchanges with other domains will not appear. Another source of complication is the ambiguity in the definition of edges in the context of a social network. In a first representation, undirected edges are drawn whenever an e-mail exchange occurs between two addresses. However, e-mails have a direction that goes from the sender to the receiver. The presence of an e-mail going from i to j does not always imply an e-mail going from j to i. This leads to a different representation in which the e-mail graph has a directed nature. Finally, we could argue that a social connection exists only between addresses which mutually exchange e-mails.

In this case a non-directed edge is present only if i has sent an e-mail to j and j also has sent an e-mail to i. All these definitions are plausible, and the different graph representations have to be chosen with respect to the particular problem or feature under scrutiny.

Ebel, Mielsch, and Bornholdt (2002) have reported the analysis of log files from the e-mail server at Kiel University. They recorded e-mail exchanges and obtained a graph with $N = 59{,}812$ vertices, including addresses outside the local domain. At first instance the undirected version of the graph has been considered, resulting in several separated clusters of about 150 vertices and a giant component of 56,969 vertices. Also for the e-mail graph the degree probability distribution is heavy tailed, obeying a power-law behavior $P(k) \sim k^{-\gamma}$ with $\gamma \simeq 1.8$ over about two decades (see Figure 7.6). The scale-free nature of the e-mail graph is confirmed by the analysis of the graph restricted only to vertices representing e-mail addresses within the Kiel domain. Also in this case the degree distribution has a power-law behavior, though its decay is slower, providing an exponent close to -1.3 (Ebel *et al.*, 2002). For the undirected e-mail graph it is also straightforward to obtain evidence concerning the small-world and clustered nature of the network. Measurements of the average shortest path length provide a value $\langle \ell \rangle = 4.95$, while the clustering coefficient turns out to be $\langle c \rangle = 0.16$, orders of magnitude larger than the value expected in a random network of comparable size and structure. Finally, Ebel *et al.* (2002) have studied the directed version of the Kiel e-mail graph. In this case, heavy-tailed behavior is obtained for both the in- and out-degree probability distributions.

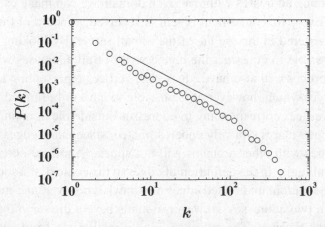

Fig. 7.6   Degree distribution of the e-mail network obtained from the Kiel e-mail server. The behavior is well approximated by a power-law form with exponent -1.8 (solid line). Data from Ebel *et al.* (2002).

The scale-free character of the e-mail network is however dependent on the definition used to draw edges and the inclusion of bulk e-mail. Guimera *et al.* (2002) studied the e-mail network built from the log files of the e-mail server of the University Rovira i Virgili. The network containing about 1,700 users has been mapped in an e-mail graph where edges were drawn only in the case of reciprocal e-mail exchanges. In addition, bulk e-mail originated from distribution lists, bulletin boards, etc., was removed. The authors of this study were indeed interested in highlighting the structure of scientific and personal relations among the users beyond the formal organization chart of the university and therefore tuned the mapping to highlight these characteristics. In this framework it is not surprising that the degree distribution of the resulting network is exponentially distributed. Considering only addresses within the domain, and discarding external and bulk e-mail traffic, generates a graph whose connectivity represents only the close social acquaintances and therefore has an extremely bounded connectivity. While this graph is not indicative with respect the study of digital infections spreading and network vulnerability to external attacks, it is particularly suited for the study of the community structure in an objective and quantitative way (Guimera *et al.*, 2002).

A different approach to characterize the social network induced by the e-mail technology has been pursued by Newman, Forrest, and Balthrop (2002). They focused on the e-mail *address book*, a file in which the user stores the e-mail addresses of regular correspondents. This should guarantee that only steady e-mail acquaintances are considered. Again, we can visualize the network as a set of vertices, the users, with directed edges from the vertex i to vertex j if the address of j is included in the address book of the user i. Newman *et al.* (2002) analyzed address book data from a large university computer system with 27,841 users. This strategy provides null information about users from the outside world, and for consistency Newman *et al.* (2002) eliminated from the network all connections to users outside the university network. The directed nature of this graph allows the calculation of the connected components structure. As already observed for other directed graphs, the usual bow-tie structure is arising. A giant strongly connected component is connected to the IN and the OUT components plus some tendrils, accounting for a giant weakly connected component which amounts to 59% of the the graph. Finally, Newman *et al.* (2002) provide also indications on the network *reciprocity*: i.e. the probability that the directed edge $i \rightarrow j$ corresponds a reversed edge $j \rightarrow i$. This probability amounts to 0.23, showing that the social connection between users often goes both ways. The directed network obtained by Newman *et al.* (2002) is composed only of users within the university domain, and it is not surprising to find that the in- and out-degree probability distributions decay faster than a power-law. This is analogous to the study carried out by Guimera *et al.* (2002), and suggests that restricting the e-mail network to the internal user community truncates the scale-free behavior observed in global studies (Ebel *et al.*, 2002).

7.3.1 The instant messaging network

While probably the most famous, the e-mail network is not the only social network due to message exchanges hosted by the Internet. For instance, in recent years *instant messaging* has become a relevant phenomenon both over the Internet and within company intranets. Instant messaging is based on a protocol that allows real time one-to-one conversation between two users. Generally speaking an instant messaging protocol uses a distributed client–server architecture. Each user (client) entering the server announces its arrival to all other users that have included its user name in their contact list, and vice-versa. Users then might decide to set up a one-to-one communication with the users present at that moment on the server or on other servers using the same instant messaging protocol.[10] Instant messaging protocols thus define a social network that can be easily identified by the users' contact list, which contains the identification of other users with whom they often communicate. In this way, one can readily define the instant messaging social network in which users are vertices with directed edges pointing to the vertices corresponding to the users contained in their contact list. Smith (2002) constructed the social network obtained by the contact lists of a French language instant messaging database. The resulting graph contains 50,158 users and 500,000 directed edges. It is no wonder that at this point the instant messaging network is found to exhibit all the characteristics of a scale-free network. Both the in- and out-degree distributions are well approximated by a power-law behavior with exponents -2.2 and -2.4, respectively. The version of the network in which edges are considered undirected confirms a degree distribution with power-law behavior and degree exponent -1.8. The analysis of the connected components' structure of the graph makes no exception to the bow-tie representation, but exhibits a massive giant strongly connected component containing 89% of the graph's vertices. This fact is due to a very high reciprocity, which shows that 82% of the contacts are bi-directional. Finally the small-world properties of the network are prompted by the very small average shortest path $\langle \ell \rangle \simeq 4$, and the high clustering coefficient $\langle c \rangle = 0.33$, four orders of magnitude larger than the one corresponding to a random graph with the same size and connectivity structure.

---

[10] Sometime instant messaging protocols support communication also among clients using different protocols.

From the previous examples, it appears that social networks on the Internet exhibit all the complex topological properties that characterize scale-free networks. In this perspective, many of the ideas and concepts put forward in the modeling of other scale-free networks might be adapted to the social context. However, the nature of the social relations has a much wider spectrum of constraints and possible mechanisms that would favor or unfavor new connections among individuals. For these reasons, no specifically devised models for the e-mail or instant messaging graphs have been put forward so far. Nevertheless, even the empirical analysis of these networks is proving to be extremely relevant in the understanding of physical and social phenomena. As we shall see in Chapter 9, the degree heterogeneity associated with the scale-free nature of these networks has profound implications in fads and computer virus spreading at the large-scale level.

7.4 Peer-to-peer and dynamic environment networks

Peer-to-peer (P2P) systems were born with the aim of exchanging files and resources between users (peers) through a direct connection among them. They have rapidly become a major social and technological phenomenon on the Internet, gaining the spotlight in view of the various issues related to copyright infringements.[11]

P2P systems are file sharing protocols that build, at the application level, a virtual network of peers with its own routing mechanism. The virtual network lives on the Internet but, obviously, is not matching the physical underlying Internet graph (see Figure 7.7). Peers are usually Internet hosts (i.e. home or office computers) which act both as client and server and communicate directly. The absence of centralized servers implies the ability of the network to cope with problems not

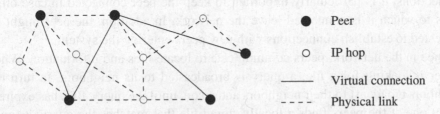

Fig. 7.7   Illustration of a P2P network. IP addresses participating to the network, the peers, establish TCP connections among them, defining a virtual network of vertices and edges. The virtual connections are not not related to the actual underlying physical or IP hop distance among vertices.

---

[11] Napster is a well-known example of P2P system used to share music on the Web. The eventual lawsuit about copyright violation has led to the collapse of Napster.

usually encountered in traditional distributed systems. The characteristic of P2P networks is, in fact, its intrinsic dynamic nature in which the large majority of computers participating in the network frequently join and leave the system and do not even possess a permanent identification.[12] It is therefore evident that the P2P network properties are emergent properties determined entirely by the collective dynamics of the interacting peers.

In order to explain how P2P systems work and give rise to the virtual network, the example of Gnutella is illuminating. This is a decentralized group membership and search protocol used for file sharing.[13] Each time a computer wants to join the network it must connect to one or more of the peers already connected by finding their address and opening a TCP connection. Since peers frequently join and leave the network, the entering computers usually try to connect with computers which are known to be almost always available. These can be obtained by lists of "good" Gnutella users, usually available on the WWW. Once in the network, the peer starts communicating with the network by broadcasting messages, i.e. sending messages to all neighbor peers with which the sender has open TCP sessions. The usual communication is made by *ping* and *pong* messages. The *ping* message announces the presence of the peer. Each time a peer receives a *ping* message, it broadcasts it to all its neighboring peers and at the same time sends back a *pong* message to the *ping* originating peer. The *pong* message contains information about the peer, such as its IP address and the number of shared files. The message broadcast is very efficient at exploring the network, but it provides a noticeable load in terms of traffic. Messages thus usually have a TTL counter,[14] so that each peer passing a message to its neighbors decrements the TTL by one. When the TTL is zero the message is not broadcasted any more. Essentially, the TTL represents the network distance to which the peer is broadcasting the message. In this way, the joining peer explores the network and discovers other peers to which it can establish additional connections. It is particularly important to keep the peer connected in case other peers to which it is connected leave the network. In addition, the peer might be requested to establish connections with new peers entering the system.

Once in the network, peers communicate to locate files and information. When a peer is looking for a file, a query is broadcasted to its neighbors. In turn the neighbors broadcast to their neighbors and so on, until the query TTL has expired. When one of the peers finds a locally stored file that matches the query, it sends

---

[12] This happens with home modem connections. In this case the computer receives from the Internet provider a temporal IP number that might change at each session.

[13] http://www.gnutella.com. Gnutella can be considered a pure P2P network. Other P2P networks, such as Napster, were maintaining a number of central servers to help the file location.

[14] The TTL (time-to-live) is measured in terms of hops on the virtual networks. The actual distance in terms of IP hops on the real network is not considered. This readily implies that P2P broadcast can generate an appreciable traffic on the physical Internet.

a response to the query originator. The response includes the information necessary to download the file and establish a direct TCP connection between the two peers. It is also important to note that a large traffic load is generated by the query. Indeed, the other peers remain unaware of the file match found and continue to flood the query through the network in any case. The expansion of P2P networks and the large traffic they generate is a major concern of the Internet community and has triggered several studies aimed at measuring the network structure, usage, and scalability of these system (Adar and Huberman, 2000; Adamic, Lukose, Puniyani, and Huberman, 2001; Saroiu, Gummadi, and Gribble, 2002; Ripeanu, Foster, and Iamnitchi, 2002).

The graph representing the Gnutella network is made by vertices representing the peers (users' computers) and edges in the place of the TCP connections among them. As we have seen for the WWW, the mapping of such a P2P network requires the development of an opportune crawler that explores the network (Saroiu *et al.*, 2002; Ripeanu *et al.*, 2002). Crawlers join the network by establishing TCP connections with a large number of peers contained in a suitable list of addresses. Then the connectivity map is obtained by broadcasting *ping* messages and recording the information contained in the *pong* replies. The possibility must be considered, however, that the network is made up of disconnected parts, since the disappearance of vertices may originate a fragmentation process. As well, different components may reconnect with the appearance of new vertices. This introduces new problems in the mapping, given that the highly dynamic environment should be extremely rapid at providing a real snapshot of the network; crawlers must thus find a trade-off between the depth and the time duration of their search (Ripeanu *et al.*, 2002).

The data collected on the Gnutella network confirm the dynamic nature of the network. This virtual network is on average steadily increasing. In a six months window (November 2000–May 2001) the largest component grew from 2,063 to 48,195 peers. The large time growth is associated with a noticeable short time variability of the network. The crawlers discovered that about 40% of the vertices leave the network in less than four hours, while only 25% are operative for more than one day (Ripeanu *et al.*, 2002). Surprisingly, despite the large number of vertices that disappear from the network, a very little fragmentation is found. In particular, the largest component includes on average more than 95% of all the vertices.

The largest component size and its stability might induce the idea of a network that, in order to cope with the vertices' dynamics, has settled in a sort of fully connected architecture. However, this is readily contradicted by the empirical evidence that shows a small and steady value of the graph's average degree, $\langle k \rangle = 3.4$. At the same time, clear evidence for small-world properties are given by a largest network diameter of 12 and an average shortest path length $\langle \ell \rangle \simeq 5$

(Ripeanu *et al.*, 2002). Finally, the degree distribution is found to exhibit a clear heavy-tailed shape that can be reasonably approximated by power-law behavior, with exponent close to -2.1 in the high degree region (Adamic, 2001; Saroiu *et al.*, 2002; Ripeanu *et al.*, 2002). The precise distribution is also changing in time, and the most recent measurements (Ripeanu *et al.*, 2002) show two separate regimes at the small and large degree values. This finding can be interpreted as a separation between occasional users, rapidly leaving the network, and Gnutella devotes, which act as the hubs and are responsible for the distribution's long tail.

The scale-free character of the Gnutella network can be reasonably explained by the network growth mechanism. New peers look first at the lists of highly available peers which already have a high degree since they are more likely to be well known. This translates into a preferential attachment mechanism that eventually leads to the scale-free degree distribution. A complete modeling of P2P networks is however a much more difficult task. Rapidly appearing and disappearing vertices should be considered on top of the usual increasing total number. The join/leave dynamics, however, is different from the usual birth/death considered in some existing models. In fact, the two processes are not uncorrelated in the sense that some of the new joining vertices represent peers that left the network sometime earlier. These vertices will likely re-establish some of the old connections and thus must have a memory of their degree and neighbors.

A first promising step towards P2P networks modeling has been recently given by Sarshar and Roychowdhury (2003). The model is defined as follows. Every time step, a new vertex is added, with m edges that are connected to previously present vertices following a linear preferential attachment. With probability c, a vertex is chosen at random and deleted from the graph, together with all its edges. Finally, all vertices that have lost edges by the random vertex deletion emanate n new edges, that are connected with the linear preferential attachment to old vertices. This model represents essentially the Barabási–Albert network with the addition of the key element of the random deletion of vertices, mimicking disconnection of peers from the P2P system. Given the scale-free nature of the initial Barabási–Albert model, the random deletion of vertices will affect mainly poorly connected vertices, while the edges subsequently removed will be with high probability connected to vertices with large degree. In their turn, those vertices will gain new connections in the rewiring process, leading to an average increase of the skewness of the distribution that is reflected in a smaller degree exponent. In fact, the continuous k approximation yields for this model a scale-free degree distribution, with degree exponent $\gamma = 1 + 2/(1 - c - 2nc)$, for $0 < n < (1 + c)/2c$ (Sarshar and Roychowdhury, 2003).

Beyond the modeling aim, the evidence for the scale-free topology of P2P graphs is fundamental to the understanding of some operative properties of these networks. The extreme stability of the network to the large amount of vertices leaving and joining the system is at first sight surprising. The leaving vertices correspond to damage of the network, that should appear fragmented while, on the contrary, data report a massive largest component. As we have seen in the previous chapter, however, this unexpected resilience to damage is a general property of networks with a scale-free degree distribution, and must not be considered a peculiarity of P2P networks. Another major issue in which the network topology enters is the P2P performance. The broadcast method finds targets quickly, but it produces a large volume of traffic on a large fraction of the network. As we mentioned earlier, in the case of the immediate success of the query, all the peers will continue the broadcast until the TTL expires. By passing the query to every vertex, the search algorithm does not take advantage of the particular connectivity pattern of the network. However, it is possible to devise different searching strategies that make use of the scale-free properties of the network and in the next chapter it will be shown how they may help to improve considerably the performance of P2P systems.

8

Searching and walking on the Internet

The problem of searching in complex networks is a very relevant issue, with a large number of practical applications. As the most obvious example of this problem we can consider the World Wide Web. It is an immense data depot that, however, acquires a practical interest and value only if it is possible to locate the particular information one is looking for. Probably, the first instance in which searching in complex networks has been considered is in the sociological context. In a now famous experiment, Milgram (1967) randomly selected people in a town in the Midwest, and asked them to send a letter to a target person living in Boston. The rule was that letters could not be sent directly but passed from person to person, only between people known on a first-name basis, until reaching the target. The unexpected result of this experiment was that the letters that arrived to the target traveled through short chains of acquaintances, with an average length of only six intermediate hops.[1] The conclusion that can be drawn from this result is not only that short paths exist between the vertices of a social network (the small-world property), but that these paths can be efficiently found in order to transmit information. The very same scenario turns out to be very relevant also for technological networks such as the Internet.

In this chapter we will review the main searching strategies that can be applied in order to retrieve information from the vertices of a network, focusing specially on the Internet and the virtual networks hosted by it. In particular, we want to provide a brief account on how the topological properties of networks might affect searching and retrieval strategies and how these strategies can effectively take adavantage of the network's connectivity structure. Obviously, the most adequate strategy turns out to depend on the amount of knowledge available on the structure of the network. The most favorable case holds when there is global knowledge of all vertices and edges, and this information is available to each vertex. In this case,

[1] From this number stems the popular expression "six degrees of separation."

the straightforward solution is to simply follow the shortest path from the starting vertex to the target vertex. Unfortunately, this favorable situation rarely takes place, and one is forced to use more expensive strategies, compatible with our knowledge of the network. As we will see, for the particular case of a scale-free network, the hierarchical organization of the vertices can be fruitfully exploited to speed up a search, by focusing first on the most connected vertices. The different strategies reviewed might be helpful for improving the performance of peer-to-peer networks and for understanding how it is possible to mine data in the WWW.

8.1 Searching strategies in networks

In order to retrieve a given item of information from a network, concerning for example the position of a particular vertex that has a certain file stored in it, as in the case of P2P networks (see Section 7.4), we can follow several searching strategies or algorithms. Searching strategies are usually described in terms of a *message passing* process that, starting from a given source vertex, passes a message to one or more of its neighbors demanding the required information, following a certain set of rules. If the selected neighbors do not have the information available, they pass the message to some of their respective neighbors, iterating the process until the information stored in the target vertex is eventually found and sent to the source.

When a complete knowledge of the network and the information stored in it is available, i.e. when every vertex knows the information stored in every other vertex and how to reach it in the minimum number of steps, one is in the best position to perform a search. Messages are passed following the sequence of vertices that compose the shortest path ℓ_{st} between the source s and the target t, and the information is retrieved in time ℓ_{st}, exchanging only ℓ_{st} messages. The average delivery time T_N to retrieve the information and the traffic measured in the number of messages M scale thus only as $\log N$ in a small-world network of size N. This situation, however, may exist only at the expense of large consumption of memory for each vertex. In the WWW, for instance, each web server should know the address and content of all the addressable web pages. In large networks this is obviously impossible. In addition, this ideal situation is altered when there are many searches going on in parallel. Assuming a finite capacity for the vertices to forward the messages that arrive to them, it is easy to see that very central vertices, routing a large fraction of the messages exchanged (vertices with a large betweenness), can easily suffer traffic congestions (for a large density of parallel searching processes), slowing down the whole network. This situation has been analyzed both theoretically and numerically on hypercubic lattices and Cayley trees by Arenas, Díaz-Guilera, and Guimerà (2001) (see also Guimerà, Arenas, Díaz-Guilera, and Giralt, 2002).

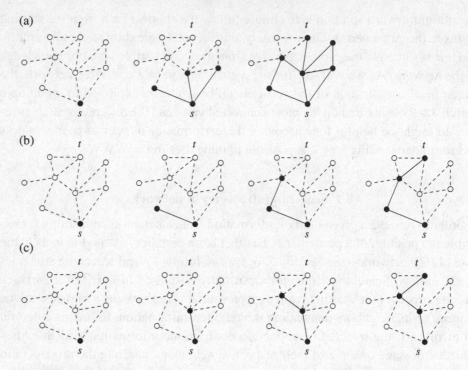

Fig. 8.1   Searching strategies to find the target vertex t, starting from the source vertex s. (a) Broadcast search, with delivery time $T = 2$. (b) Random walk search, $T = 3$. (c) Degree-biased search, $T = 3$. The broadcast search finds the target faster, but it involves a larger number of exchanged messages. Figure adapted from Kim, Yoon, Han, and Jeong (2002).

In the opposite case of absence of any knowledge on the information structure of the network, i.e. when no vertex knows the exact position of the required information until it is found, the simplest strategy one can follow is a *broadcast search* (see Figure 8.1(a)). In a broadcast search the source vertex sends a message to *all* its nearest neighbors, which if they do not have the requested item, send on their turn a message to all their neighbors, excluding the source vertex. This process is repeated a certain number of times, since usually a time-to-live (TTL) is assigned to the search. After a message has been forwarded from neighbor to neighbor an amount of time equal to the TTL, it is discarded. When a vertex containing the sought-after information is reached by the query, it sends to the source vertex a message to establish the exchange of information. Usually, however, the target vertex cannot stop the broadcast search going on in the rest of the network, but at the cost of a large number of messages. In this sense, a broadcast search is essentially equivalent to the breadth-first algorithm used to compute the shortest path length between any two vertices, and corresponds to a complete sweeping of all

the vertices within a TTL hop distance from the query source. This search strart-
egy is also very rapid, since it proceeds in parallel, and in a few time steps all the
network can be explored. Indeed, it is easy to see that the delivery time needed to
find the target vertex t starting from the source s, measured as the number of intermediate edges traversed (one every time step), is equal to the shortest path length
ℓ_{st}. In terms of the average delivery time T_N as a function of the graph size N,
a broadcast search is thus very efficient for small-world networks, since we have
that T_N scales as the average shortest path length $\langle \ell \rangle$, and therefore grows only as
the logarithm of N or slower (see Section 5.6). However, it has the severe drawback of requiring a very large amount of traffic, since a large fraction of vertices are
visited and forced to exchange messages among them. In fact, for the case of a generalized random graph, we have seen in Section 5.1.2 that the average number of
neighbors at distance d of any given vertex grows exponentially with d. Therefore,
in a small-world network, even for searches with small TTL that deliver results in
a short time, one is forcing almost all vertices to exchange messages, imposing an
average amount of traffic that increases linearly with the size of the network.

In general, however, intermediate cases in which there is some amount of local
information available about the structure of the network may possibly be devised.
In this situation, it is possible to develop more economical searching strategies
that do not put such a severe strain on the traffic exchanged. For instance, let us
assume a situation in which each vertex knows only about the information stored
in each of its nearest neighbors. In this case, the most naïve economical strategy
is the *random walk search*, Figure 8.1(b), in which the source vertex sends *one*
message to a randomly selected nearest neighbor. If that vertex has the information requested, it retrieves it; otherwise, it sends a message to one of its nearest
neighbors, avoiding sending it back to the source. This process is iterated until a
neighbor of the vertex in possession of the information item requested receives a
message, in which case it is forwarded directly to the target vertex. It is easy to see
that this strategy is less efficient than a broadcast search in terms of the delivery
time. In fact, a random search tends to backtrack its own path, delivering messages
to vertices that have already passed them. Therefore, the delivery time tends to be
longer than the average shortest path length yielded by a broadcast search. In particular, computer simulations performed on a scale-free generalized random graph
(see Section 5.1.2) with degree exponent $\gamma = 2.1$, equal to the value observed in
real P2P networks, yield the result $T_N \sim N^{0.79}$ (Adamic *et al.*, 2001). As we can
see, in terms of the time needed to locate a given vertex, the random walk search
fares worse than a broadcast search in a scale-free network. However, since at each
time step only one message is sent, the average total traffic in the network is equal
to T_N, and therefore scales sublinearly with N, better than the linear growth typical
of a broadcast search.

8.1.1 Using the shortcuts

When dealing specifically with small-world networks, it is possible to devise more efficient searching strategies levering on this very property. As we have seen in previous chapters, the main characteristic of small-world networks is the presence of shortcuts, that act as bridges between far away parts of the network and lead to a very noticeable reduction in the average diameter of the graph. In performing a search in this kind of network, it is therefore reasonable to assume that it should be possible to use those shortcuts to one's advantage. While this intuition seems to be correct, there is actually a considerable difference between knowing that there is a shortcut and actually finding and using it to reach a particular vertex (Watts, 2003). If the shortcuts are placed between randomly selected vertices, as in the Watts–Strogatz model (Chapter 5), then every vertex has the same probability of being connected to any place in the network and a search strategy based on an intelligent use of shortcuts is doomed to fail, since it is impossible to find the right shortcut leading to the desired point. The conclusion is that, in order to be useful, shortcuts must encode some information about the underlying structure of the network.

The realization of this fact led Kleinberg (2000) to study the conditions under which it is possible to perform an efficient search in a Watts–Strogatz-like small-world model. In order to do so, he considered a D-dimensional hypercubic lattice in which a certain fraction of shortcuts are added. The probability of connecting two vertices at a geographical distance r decreases as $p(r) \sim r^{-\alpha}$, where the exponent α is the one controlling the behavior of the model. On this graph, the following directed "greedy" searching algorithm was studied: each vertex passes a message to its nearest neighbor that is the closest in geographical distance to the target of the search. Clearly, this is a directed strategy, in which the message tries to get as close as possible to the target at each step, implying a large amount of global knowledge of the structure of the networks, i.e. every vertex knows where the target is, and which one of its neighbors is closer to it.[2] By means of theoretical arguments and numerical simulations, Kleinberg (2000) concluded that the average delivery time experiences a transition at the particular value $\alpha = D$, see Figure 8.2. Whenever the exponent α is different from the lattice dimensionality D, the average delivery time is large, growing as a power-law of the system size.[3] Exactly at $\alpha = D$, however, T_N reaches its minimum value, being bounded by a polinomyal of $\log N$.

The interpretation of this result is the following (Watts, 2003). For $\alpha > D$ the number of long-range shortcuts is very small, and essentially they do not exist. In

---

[2] However, there is not a complete knowledge of the network, i.e. the vertices do not know the shortest path to the target, only the next hop that will place the message closer to it.

[3] For $D = 2$, a lower bound $T_N \geq N^{\beta(\alpha)}$ can be provided, with $\beta(\alpha) = (2 - \alpha)/3$ for $0 \leq \alpha < 2$ and $\beta(\alpha) = (\alpha - 2)/(\alpha - 1)$ for $\alpha > 2$ (Kleinberg, 2000).

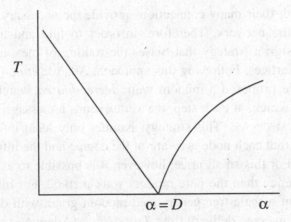

Fig. 8.2   Average delivery time as a function of the exponent α for a small-world model with local shortcuts. Figure adapted from Kleinberg (2000).

the opposite case, $\alpha < D$, long-range shortcuts are present in the system, but they cannot be efficiently found by the searching algorithm. In particular, the Watts–Strogatz model belongs to this regime. Only at the critical point $\alpha = D$, when vertices possess on average the same number of shortcuts at *all length scales*, can the network be efficiently navigated using the bridges to gap long jumps in the direction of the target.

The conclusion of Kleinberg (2000) seems to imply that only very special kinds of small-world networks can be efficiently navigated, that is, only those with the right distribution of long-range shortcuts. The key to understanding how real networks can be searchable[4] is to realize that more than one distance can be defined on them, attending to other criteria, such as affinity on different subjects. Based on this premise, Watts, Dodds, and Newman (2002) (see also Kleinberg, 2002). Proposed a model of searching in a network in which distance can be defined along several dimensions, concluding that the domain in which the network is searchable is much broader than a single critical point, as in the model of Kleinberg (2000).

8.1.2 Degree-biased search in scale-free networks

While an efficient exploitation of the "greedy" algorithm we have seen in the previous section presupposes an extensive knowledge of the network that is usually absent for the particular case of distributed scale-free networks, it is still possible to fruitfully pursue the intuition of using the shortcuts. As we have seen in Chapters 4 and 5, the small-world character of scale-free networks is due to the presence of

---

[4] At least social networks, such as the one considered in Milgram's experiment.

the hubs, that with their many connections provide the necessary bridges among distant parts of the network. Therefore, in order to find and use those short-cuts, one can design a strategy that biases the routing of messages towards the *most connected* vertices. Following this approach, Adamic *et al.* (2001) and Kim *et al.* (2002) have proposed a random walk *degree-biased* searching algorithm, Figure 8.1(c), in which, at each step, the vertex sends a message to the neighbor that has the largest degree. This strategy assumes only local information in the network, namely that each node is aware of the degree and the information stored in its neighbors. For this small price, however, it is possible to explore the graph with larger efficiency than the pure random walk method. For instance, numeri-cal simulations on a scale-free generalized random graph with degree exponent $\gamma = 2.1$ yield an average delivery time $T_N \sim N^{0.70}$ (Adamic *et al.*, 2001), i.e. a degree-biased algorithm can retrieve information in a scale-free network faster than a simple random walk search. This numerical evidence is supported by the analytical approach of Adamic *et al.* (2001), focusing on the dependence of the average delivery time with the cut-off of the network degree distribution. The rea-son for the success of a degree-biased search is easy to understand: if every element has knowledge of the contents of its immediate neighbors, the natural way to in-spect the maximum number of vertices at each time step is to start asking the most connected vertex and go down in the connectivity hierarchy. Indeed, it is possi-ble to see that, starting from a randomly selected vertex and passing a message to the neighbor with the largest degree, in a few steps the message is located on the most connected vertex, from where it more or less follows a decreasing degree se-quence (Adamic *et al.*, 2001). In this respect, the rule of deterministically sending the message to the neighbor with the largest degree turns out to be crucial. Any amount of randomness in this selection, as for example sending the message to the neighbors with a probability proportional to their degree, yields a much longer average delivery time (Kim *et al.*, 2002).

Apart from locating information, searching strategies are useful to find paths be-tween different vertices, paths that can subsequently be reused to contact again the same vertex (Kim *et al.*, 2002; Adamic *et al.*, 2003). In order to construct a short-est path, of length ℓ'_{ij}, from the path followed by the message passing process, one only needs to remove loops and backwards steps. The average path lenght $\langle \ell' \rangle$ ob-tained by this process is longer that the average shortest path length, see Figure 8.3. This fact is confirmed by numerical simulations of the random walk search of the Barabási–Albert network, that yields $\langle \ell' \rangle \sim N^{0.51}$ (Kim *et al.*, 2002). However, the degree-biased search on the same kind of network produces an average path scal-ing logarithmically with N, in the same fashion as $\langle \ell \rangle$. A degree-biased random search is therefore effective also for estimating the average shortest path length of a scale-free network. Its performance, however, is handicapped by the average cost

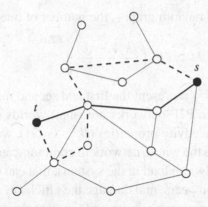

Fig. 8.3 Difference between the shortest path between vertices s and t (continuous thick line), with length $\ell_{st} = 3$, and the path found with the random walk search strategy (dashed thick line), of length $\ell'_{st} = 6$.

of finding a path (the average delivery time), that in the Barabási–Albert network scales linearly with N (Adamic *et al.*, 2003).

The conclusion of this section is that the hierarchical structure of scale-free networks can be efficiently exploited by means of degree-biased strategies in order to reduce the average cost of a search. Visiting in the first place the vertices with the largest degree allows us to explore a large fraction of the network in the very first steps of the search process. However, the same strategy imposes a drawback in the process, in the sense that, after the initial period in which new vertices are visited, the search tends to trace back its steps and visit more often the already explored high degree vertices. Therefore, vertices with low degree can be relatively difficult to locate, yielding a heavy tailed distribution of delivery times that is responsible for the final scaling of T_N (Adamic *et al.*, 2001).

8.2 Improving the performance of peer-to-peer networks

Peer-to-peer (P2P) systems are another example of networks in which searching efficiency is a major issue. As we have seen in Section 7.4, the searching mechanism implemented in P2P networks is a broadcast search limited to a radius TTL, the time-to-live assigned to the queries. While this procedure is able to locate the information present in a radius TTL in at most TTL hops between peers, it is obvious that it implies that many nodes in the network are forced to exchange messages. In fact, since the peers do not know when the requested file is found, and keep sending the query until the TTL is expired, we can see that the total number of sent queries M is approximately given by the number of vertices within the TTL distance measured in number of hops. By using the result of Eq. (5.12) for generalized uncorrelated random graphs, the number of queries is given by

$$M \simeq \langle k \rangle \left(\frac{\langle k^2 \rangle - \langle k \rangle}{\langle k \rangle} \right)^{TTL-1}. \tag{8.1}$$

where as usual $\langle k \rangle$ and $\langle k^2 \rangle$ represent the first and second moments of the network degree distribution. Since P2P networks are small-worlds ($\langle \ell \rangle \sim 5$, Section 7.4) with heterogeneous connectivity properties ($\langle k^2 \rangle \gg \langle k \rangle$), we see that even a small value of TTL might force the whole network to communicate for every query. This fact imposes a large bandwidth load in the systems, that can eventually saturate the TCP connections between peers, and damage the efficiency of the network.

Several alternatives are possible in order to reduce the traffic load in P2P networks, and consequently increase their performance. The most simple is the creation of a centralized repository of information, containing the names of all the files available and the corresponding addresses in which they can be found. This philosophy, adopted in the original Napster, has the drawback of being expensive (powerful servers are needed in order to store and manage efficiently this information) and, worst of all, liable to copyright infringements. Discarding this dangerous solution, searches can be improved within the actual P2P protocol by using an *iterative deepening* (Russel and Norvig, 1995) algorithm, as proposed by Yang and Garcia-Molina (2002). In this method, broadcast queries are sent with an increasing TTL value. Experiments in actual Gnutella networks show that a large fraction of bandwidth can be saved this way (up to a 72% with respect to the standard method), reducing as well the processing cost of the search.

It is also possible to apply the concepts developed in Section 8.1.2 in order to implement a radically different searching protocol that takes advantage of the scale-free nature of P2P systems (Adamic *et al.*, 2001). As we have seen, degree-biased strategies are able to locate files in times that scale sublineraly with the size of the network, but, most importantly, require traffic that is also sublinear in N. The trade-off between a longer search and a less-congested network could probably lead to better global efficiency in terms of real delivery times. Computer simulation shows that this is indeed the case. For example, studies of generalized random graphs with degree exponent $\gamma = 2.1$ show the time needed to explore half of the network[5] with a degree-biased strategy grows only as $N^{0.24}$ (Adamic *et al.*, 2001). However, simulations on an actual Gnutella network of size 700 show that half of the vertices can be searched in eight or less hops. This new strategy, however efficient, is nevertheless not free from a few inconveniences. First of all, it would imply a complete rewriting of the Gnutella searching protocol, allowing each peer

---

[5] Since in P2P networks more than a vertex is probably storing the required information, this is a more significative measure than the average delivery time from a single target.

to store additional information about the contents of its nearest neighbors.[6] But, more importantly, a degree-biased search implies that almost all queries are eventually routed to the most connected peers, which will experience the largest load of processing and bandwidth. Nevertheless, since the Gnutella protocol restricts the number of connections that peers are willing to accept, it is probably safe to assume that the vertices with the largest degree are also the ones with the largest computing resources and bandwidth, and therefore they should be able to cope with the increased load placed upon them.

8.3 Searching on the Web

As we have seen in Chapter 7, the World Wide Web is a virtual nework, hosted by the physical Internet, that offers to the casual web user the largest amount of information ever collected. It has been compared with a "15 billion word encyclopedia" (Barrie and Presti, 1996), and this metaphor is probably an underestimation nowadays, given the pace at which the Web is increasing.[7] In order to be able to find the desired information in such a gigantic repository, the development of specifically designed searching tools capable to deal with the size of this system becomes mandatory.

8.3.1 Static search engines

Many public *search engines* are available on the Web, such as "Google",[8] "Yahoo!"[9] or "AltaVista",[10] to help users find information on the WWW. These search engines (Marendy, 2001) rely on a static index of words found in the Web, that is usually collected by automatic programs called *Web crawlers* or *spiders* (see Section 7.1.2).[11] The crawlers follow a list of links provided by a central server or follow recursively the links they find in the pages that they visit, according to a certain set of searching instructions. When a crawler finds a new web page in its search, it stores the data it contains and sends it to a central server. Afterwards, it follows the links present in the page to reach new web sites. In this sense, the

---

[6] The extra information that each vertex should store might not eventually be very large. For example, Yang and Garcia-Molina (2002) estimate that storing the information of a small neighborhood would imply a memory overhead of the order of 50 Kilobytes per peer.

[7] In 1999 a rate of growth close to one million new web pages per day was estimated by Chakrabarti, Dom, Gibson, Kleinberg, Kumar, Raghavan, Rajagopalan and Tomkins (1999).

[8] http://www.google.com.   [9] http://www.yahoo.com.   [10] http://www.altavista.com.

[11] Alternative search tools are the "Web Directories," that provide a hierarchical classification of web pages that are collected (sometimes by direct submission from the authors), scrutinized, and reviewed by a board of editors. This human-powered directories aim at providing classified information of contrasted quality and relevance. An example of this kind of search tools are Yahoo, http://www.yahoo.com, or the Open Directory Project, http://dmoz.org.

strategy followed by web crawlers is a simple broadcast search. Web crawlings are repeated at periodic time intervals, to keep the index updated from new pages and links, and to delete old or obsolete directions. The information retrieved by the crawlers is analized and used to create the index. The index stores information relative to the words present in the web pages found, such as their position and presentation, forming a database relating those words with the relevant hyperlinks to reach the pages in which they appear, plus the hyperlinks present in the pages themselves.

The final element in a search engine is the user interface, a search software that accepts as an input words typed by the user, explores the index, and returns the web pages found that contain the text introduced by the end user, and that are considered more relevant. In this process the *ranking* of the pages returned is most important, i.e. the order in which they are presented after the query. Obviously, nobody is willing to visit dozens of uninteresting pages before discovering the one that contains the particular information that is seeked. Therefore, the more relevant are the first pages returned, the more succesful and popular will be the search engine. The search engines available in the market make use of different ranking methods, based on several heuristics for the location and frequency of the words found in the index. Traditionally, these heuristics combine information about the position of the words in the page (the words in the HTML title or close to the top of the page are deemed more important than those in the bottom), the length of the pages and the meaning of the words they contain, the level of the directory in which the page is located, etc.

In this respect, the PageRank algorithm proposed by Brin and Page (1998), on which the search engine "Google" is based, came as a real breakthrough. PageRank uses a graph theoretical analysis of the in-degree distribution of the pages in the index, combined with some heuristics based on the text disposition. In order to do so, the index has to store not only the words present in the pages crawled, but also the structure of the hyperlinks between pages, i.e. the graph representantion of the Web. The main idea is to assign the page's relevance or popularity on the basis of the number of edges that point toward them, a concept we have seen at work in the development of Web models, being the main motivation for the preferential attachment mechanism (see Section 7.2). In practice, a rank $P_R(i)$ is assigned to each page i, that is computed by means of the recursion relation (Brin and Page, 1998)

$$P_R(i) = (1 - d) + d \sum_j A_{ji} P_R(j)/k_{out,j}, \tag{8.2}$$

where A_{ij} is the adjacency matrix of the Web graph and $k_{out,j}$ is the out-degree of vertex j. The rank assigned to each web page from Eq. (8.2) can be interpreted as the probability that a random web surfer, that wanders in the Web by clicking the links he finds, visits the page i. In this interpretation, the *damping parameter* d in Eq. (8.2) represents the probability that the random surfer gets bored, stops following links, and proceeds to visit a randomly selected web page. Iterating the PageRank algorithm a sufficient number of times, a probability $P_R(i)$ is assigned to each page, that can afterwards be used to classify very quickly the results of the search.

An alternative method to improve the ranking of search engines by considering both the in- and the out-degree distribution of the Web graph has been proposed by Kleinberg (1998). This method relies on the distinction between *authorities* and *hubs*. Authorities are Web pages that can be considered the most relevant source of information about a given topic. Given the large amount of knowledge that this kind of page encodes, it is natural to assume that they have a large number of incoming links. Hubs, however, are pages dealing with a given topic, which are not authorities themselves but which contain a large number of outgoing links, pointing to related authorities. In this situation, the set of hubs and authorities on a topic form a bipartite clique (see Section 7.2), in which all hubs point to all authorities. Therefore, focusing on the detection of bipartite cliques, it should be possible to identify which are those authorities and rank them in the highest position. Following this approach, Kleinberg (1998) proposed the Hyperlink-Induced Topic Search (HITS) algorithm, which has been the seed for several variations and improvements (Marendy, 2001).

The search engine technology makes it possible to find in most cases a very large number of pages (hundreds or more) of content related to the initial query introduced, a fact that could induce to think that quite a good coverage of the WWW has been achieved. This impression, however, is not backed up by the empirical evidence. Indeed, Lawrence and Giles (1999) estimated that none of the six major public search engines cover individually more than a 16% of the estimated Web size, the largest one searching on 128 million pages out of the 8×10^8 documents of the publicly indexable Web,[12] while collectively they cover a 42% of the total.[13] In spite of their continued struggle to provide increased coverage, the sheer size of the Web and its exponential growth seems to preclude the possibility of a complete mapping of this network. As a matter of fact, the performance of the search engines is decreasing with time, since in 1998 it was estimated a maximum coverage of any single engine as 34% (Lawrence and Giles, 1998).

---

[12] These values corresponding to 1999 might be at present outdated. In fact, as on March 2003, "Google" claims to be able to search in more than 3×10^9 web pages, that with the estimated coverage of a 7.8% for this search engine (Lawrence and Giles, 1999), yields a Web of size close to 4×10^{10}.

[13] Therefore, a simple way to improve the performance of a web search is to use a meta search engine, such as http://www.monstercrawler.com, that combines the results of many single engines.

The size of the Web is not the only fact hindering the performance of search engines. The very topology of the WWW poses an intrinsic hindrance, much too difficult to deal with (Barabási, 2002). As we have seen in Section 7.1.2, the WWW is a directed network, a fact that has a deep impact in the component structure of the Web graph. Looking at Figure 7.1 we can easily see that, by following the links present in each web page,[14] and starting from a page belonging to the giant strongly connected component (GSCC), it is possible to explore the full GSCC plus the OUT component. Therefore, these two components of the Web, that amount to roughly 50% of the whole network, are fully available to crawlers. Whenever a new document appears with at least one incoming link from the GSCC or the OUT components, it will be eventually found and indexed. However, in order to reach a new document located in the IN component or the tendrils, i.e. a web page that points to the GSCC but is not pointed by it, it is necessary to start the crawl from these sections of the Web, and hope to find the necessary chain of links leading to it. To provide a remedy to this situation, many search engines allow the submission of web pages from their authors, in order to start from them new crawls that could unveil new sectors of that hidden WWW.

8.3.2 Searching the Web in real time

In opposition to traditional web search engines, that rely on a static index of words that is periodically updated by extensive Web crawlings, the possibility has been proposed recently to use adaptive multiagent systems capable of performing in real time online search of the Web at the moment of the user's query (Menczer, 2003; Menczer and Belew, 2000; Chakrabarti *et al.*, 1999; Aggarwal *et al.*, 2001; Cho *et al.*, 2000). The example of this new paradigm is the InfoSpiders[15] model, developed by Menczer and Belew (2000). InfoSpiders is a system composed by a population of agents that crawl the Web visiting pages. The agents are able to dynamically adapt to the environment of the pages they visit. A query with this system starts with an input of keywords provided by the user, plus a set of Web pages relevant to the query in question, that can be extracted from a traditional search engine. An agent is placed in each of those pages and proceeds to examine the contents of the page in order to estimate the relative relevance of the documents to which a given page points. The information collected by the agent is transmitted to a neural network that decides which are the links with a higher relevance. Finally, the state of the agent is updated, deciding whether it dies, follows a link, or spawns additional agents to follow more links in case a promising area of information is found.

---

[14] As crawlers do in their broadcast search strategy.   [15] http://myspiders.biz.uiowa.edu.

The key element of InfoSpiders is the capability to adapt and learn during the crawl. In some sense, therefore, it behaves like a human performing "smart" web surfing, following the links in the pages found only after analyzing its probable content and deciding the best direction to find the required information. From this point of view, these systems can be encompassed in the class of greedy algorithms described in Section 8.1.1, in which each visited vertex is increasingly closer in geographical distance to the target vertex. In fact, it can be shown that, using lexical distance in place of geographical distance, the Web's topology satisfies the necessary and sufficient conditions postulated by Kleinberg (2000) for efficient navigation (Menczer, 2002).

Despite the fact that they are still limited by the size and directed topology of the Web, adaptive multiagent systems represent an improvement over traditional static search engines. First of all, they avoid the large load imposed on the Internet by the periodic "blind" crawlings that are needed to update static word indexes. They also are able to provide fresher data than traditional search engines; since they perform real time searches, they can find very recent documents.[16] Additionally, it can be seen that on average, multiagent systems can deliver pages more relevant to the query in question than a traditional system (Menczer, 2003). InfoSpiders thus opens new and promising perspectives as a complementary approach to traditional search engines, in a symbiosis in which both systems can take advantage of the benefits of the other, in order to provide faster, reliable, and more accurate searching tools to explore the immensity of the World Wide Web.

---

[16] In comparison, it takes on average 180 days for a new web page to be indexed in a traditional engine (Lawrence and Giles, 1999).

9

Epidemics in the Internet

The Internet is a technological infrastructure aimed at favoring data exchange and reachability. The World Wide Web can be used to extract information from distant places with a few mouse clicks, and Internet protocols forward messages to far away computers, efficiently routing them along the intricate network fabric. This extreme efficiency, however, can also work in favor of negative purposes, such as the spreading of computer viruses. Computer viruses have a long history, dating from the 1980s and before, becoming newly and sadly famous after each new bug attack, which eventually causes losses worth millions of dollars in computer equipment and downtime (Suplee, 2000). Their ever-increasing threat has therefore stimulated a growing interest in the scientific community and the economic world, translated in this latter case into the antivirus software business, moving millions of dollars worldwide every year.

Computer virus studies have been carried out for long time, based mainly on an analogy with biological epidemiology (Murray, 1988). In particular most studies have focused on the statistical epidemiology approach, aimed at modeling and forecasting the global incidence and evolution of computer virus epidemics in the computer world. The final goal of this approach is the development of safety and control policies at the large-scale level for the protection of the Internet. Puzzling enough, however, is the observed behavior of computer viruses *in the wild*,[1] which exhibit peculiar characteristics that are difficult to explain in the usual epidemic spreading framework. In particular, digital viruses appear to easily reach long-lasting, almost endemic, steady states, corresponding to generalized very long lifetimes, indistinctive of the viral strain. In addition, massive immunization campaigns do not eradicate the viruses with the expected efficiency. The key to understanding all these features is the particular background in which computer virus activity carries on. While classical epidemiological models consider viruses

---

[1] That is, viruses found by actual users in the real world. See the Web page http://www.wildlist.org.

propagating on the vertices of regular lattices or random networks with rather homogeneous degree distributions, computer viruses dwell in a digital world (the physical Internet, the WWW, the e-mail network), which, due to its heterogeneous nature, has intrinsically large degree fluctuations. The introduction of this new element yields a theoretical scenario in which all viruses, irrespective of their virulence, have a chance to pervade the whole system. In this sense, the Internet, and in general all scale-free networks, are very weak in the face of infections.

In this chapter we provide the general framework of epidemic modeling in complex networks, showing how the introduction of degree fluctuations leads to a rationalization of empirical data from computer virus epidemics. Finally, we discuss to what extent the protection of the Internet and cyber-communities, defined in the context of immunization policies designed to effectively reduce or prevent the large-scale spreading of computer viruses, must take into account the heterogeneous nature of these networks.

9.1 Computer viruses and worms

The first kind of bug affecting the Internet can be identified in spontaneus error propagation mechanisms at the software level. For instance, Bellovin (1993) described the DNS cache corruption spreading as a *natural computer virus* proliferating on the Internet. As we have seen in Chapter 2, computers on the Internet rely upon Domain Name System (DNS) servers to translate IP addresses into computer names and vice-versa. In turn, DNS servers communicate with each other to share and update this information. Translation tables are "cached" and eventually transmitted to the other DNS peers. If any portion of this cache is corrupted, the DNS server will provide incorrect addresses not only to requesting computers but to DNS peers as well, propagating the error. At the same time, any DNS server can get "cured" by updating with an error-free DNS peer or by manual intervention. The same kind of processes can occur with routing tables exchanged by routers. Error propagation occurring on routers and servers that are physically connected can thus be considered as an example of an epidemic process, in which the corruption (bug) is transmitted from infected to healthy individuals.

From a more familiar point of view, however, computer viruses are usually referred to as little programs that can reproduce themselves by infecting other programs and computers (Harley, Slade, Harley, Spafford, and Gattiker, 2001; Kephart, Sorkin, Chess, and White, 1997). Unfortunately, apart from reproducing themselves, computer viruses perform other threatening tasks, which range from flashing innocuous messages on the computer screen to seriously corrupting data stored in the hard drive. These deleterious effects render most computer viruses as dangerous as their biological homonyms, and explain the interest, both commercial and scientific, which has arisen around their study (Kephart *et al.*, 1997). Leaving aside academic experiments on the possibility of creating parasitic programs on computers, the first virus found *in the wild* dates from 1981, a quite innocuous bug of the Apple II computer. The first virus capable of infécting a PC was the *Brain* virus, developed in Pakistan in 1986. From this humble origin, digital viruses have risen to constitute a real economic and technological threat. In the year 2000 it was estimated that there were more than 48,000 identified different viruses, of which more that 10,000 were spreading *in the wild*. Scaringly enough, new viruses are being discovered at a rate of more than ten every month.

Schematically, the mechanism of infection by computer viruses is as follows. When the virus is active inside the computer, it is able to copy itself, by different ways, into the code of other, clean, programs. When the newly infected program is run into another computer, the code of the virus is executed first, becoming active and being able to infect other programs. Starting from primitive mechanisms, however, the skills of virus programmers have led to the development of very sophisticated pieces of code, capable of hiding themselves from the scrutiny of software engineers. In other words, cyber viruses have evolved in time (driven of course by their programmers' skills), adopting new strategies that take advantage of the different weak points of computers and software. According to their different infection and transmission mechanisms, computer viruses are classified into three basic categories, or *strains* (Kephart *et al.*, 1997):

1. *File viruses* infect application programs. When executing an infected application, the virus is executed first and is independently installed into the computer's random access memory (RAM). Whenever a new, clean application is subsequently run, the virus copies itself into the executable file of the application, infecting it. File viruses spread mainly via the sharing of applications.
2. *Boot-sector viruses* infect the boot sector of floppy disks and hard drives, a portion of the disk containing a small program in charge of loading the operating system of the computer. When the computer is started, the code of the virus is read from the boot sector, and becomes installed in the memory, ready to infect new floppy disks inserted into the computer. Boot-sector viruses spread usually via the sharing of infected floppies.
3. *Macro viruses* are independent of the hardware's architecture and operating systems, and infect data files, such as documents produced with word processors or data sheets. They are coded using the *macro* instructions appended in the document, instructions used to perform a set of automatic tasks, such as formatting the documents, or typing long sequences of characters. Their transmission is via the sharing of infected documents, that are themselves transmitted by different ways.

Very soon, however, a wealth of hybrid viruses, developed in response to the ever-increasing deployment of new and more efficient antivirus software, made their appearance. These mix the properties of the main strains in more harmful combinations. For instance, they can infect at the same time the boot sector and application programs (*multipartite viruses*), boosting in this way their infection efficiency.

The digital virus world experienced a major upheaval in the late 1990s with the appearance of the nowadays dominant and most aggressive type of cyber organisms, the computer *worms* family. The first representative of this family is the *Melissa* virus, discovered in 1999. Worms are cyber viruses infecting the computer with mechanisms similar to regular biological viruses and making particularly effective use of e-mail for infecting new computers. In fact, by using the macro instructions of some commercial e-mail software applications, worms are capable of sending themselves to all the electronic addresses found in the address book of the person receiving the infected mail. This possibility renders worms very effective viruses, especially in terms of the velocity at which they can propagate, starting from a small core of infected computers. As an example of the enormous efficiency of worms, one of the most virulent of them, the *I-love-you* bug, was able to infect more than 78 million computers worldwide in scarcely four days (Buchanan, 2002).

As if the above scenario were not worrying enough, the taxonomy of cyber plagues is constantly increasing. Worms nowadays have the capability of spreading over different networks and making use of different protocols. The last in order of appearance are *active worms*, that do not need any user intervention to propagate, detecting bugs in operating systems and guessing IP addresses to attack (Moore, Shannon, and Brown, 2002; Staniford, Paxson, and Weaver, 2002). The velocity of contagion of these worms is enormous and many researchers and practitioners in the field are so worried about the cyber virus threat that they are pushing for the development of "centers for disease control," analogous to those operating in the case of biological viruses (Staniford *et al.*, 2002).

9.1.1 *Statistical data on computer virus spreading*

The properties of computer virus spreading have been analyzed by several authors, in close analogy with the classical epidemiology of biological diseases (Bailey, 1975; Anderson and May, 1992; Murray, 1993; Diekmann and Heesterbeek, 2000). Indeed, in many cases it is possible to have a one-to-one mapping of biological parameters with those of the cyber world (Kephart, White, and Chess, 1993; Kephart *et al.*, 1997; Aron, O'Leary, Gove, Azadegan, and Schneider, 2002). In both cases, the transmission of the virus is due to contact or interaction between an infected and a susceptible individual. Immunization is present in both cases, referring, in the computer world, to anti-virus scanning, operating systems updates, and other forms of staying alert. Also the medium for transmission is in both cases

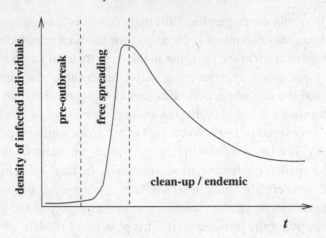

Fig. 9.1 Stages of an epidemic outbreak. The clean-up region can define a long-standing stationary number of infections resulting in an endemic state. Figure adapted from Schwartz and Billings (2003).

a network. In biology there are human interaction webs, sexual webs, food webs, etc., while in the cyber world there are the Internet, the e-mail, and other social networks, such as P2P systems. Finally the same extra features, such as seasonal or time effects, are generally encountered in both contexts.[2]

In this context, statistical studies of computer epidemic outbreaks have focused on quantities which are customarily used in the characterization of biological diseases. In particular, measurements concern virus *prevalence*, defined as the average fraction of computers infected with respect to the total number of computers present.[3] In more detail, the behavior over time of the number of infected individuals yields a general characterization of the epidemic's outbreak intensity and time scale. In general, real data show that large-scale epidemic outbreaks follow a three stage sequential scheme depicted in Figure 9.1 (Schwartz and Billings, 2003). First, there is a pre-epidemic stage in which the new virus is formulated and inoculated into the population. A second stage refers to a *free-spreading* phase in which the virus does not find any resistance because of the lack of any scanning or "curing" antivirus software. At a third stage, defensive software, system patches, and recovery procedures are developed and the virus enters a *clean-up* stage in which the number of infected individuals is decreasing.[4]

[2] There are however some basic differences between biological and computer viruses that are worth mentioning. First, biological epidemics grow on a slower time scale. Secondly, biological individuals are self-reacting against viruses with their immune system, while computers are not.

[3] The density of infected individuals in the population at time t is usually referred to as *incidence*. The prevalence refers to the total number of individuals since the beginning of the epidemic outbreak. In the case of endemic states the prevalence refers also to the stationary state value of density of infected individuals.

[4] This general picture may however find exceptions. In particular, the most recent active worms exhibit more complicated temporal patterns (Moore *et al.*, 2002; Staniford *et al.*, 2002).

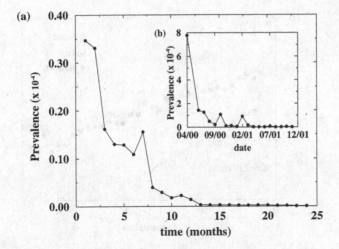

Fig. 9.2 (a) Average prevalence of computer worms as a function of the age of the virus, measured in months. (b) Time evolution of the prevalence of the *vbs/loveletter* virus, a variant of the *I-love-you* bug. In the plot it is not possible to report the very fast increase of the free spreading phase that is smaller than the monthly data time resolution.

Data collection is not an easy task in computer epidemiology. The early free-spreading phase is very fast and it is difficult to obtain data in experimentally controlled situations. There are also privacy issues concerning companies and individuals and often a lack of knowledge on a global scale since data are usually circumscribed to virus prevalence in some specific domain. From various studies (Kephart and White, 1993; Kephart *et al.*, 1993; Kephart *et al.*, 1997; Pastor-Satorras and Vespignani, 2001b; Staniford *et al.*, 2002), however, it is possible to draw some general empirical evidence concerning the clean-up stage and overall virus lifetime.

First evidence indicates that the clean-up stage is usually very long, eventually settling in a long-lasting stationary state that can be considered as an *endemic* state. Noticeably, this state goes along with a *very small* average prevalence, that can be of the order of 1 out of every 10,000 computers or less. This evidence as been observed both in viruses and worms (Kephart and White, 1993; Kephart *et al.*, 1993). As an illustration of this feature, Figure 9.2(a) shows the incidence of computer worm outbreaks a time t after the first observation. Data are averaged over all the outbreaks over the period from April 2000 to May 2002 as estimated from the monthly data summary provided by *MessageLabs*,[5] a managed service provider specializing in e-mail security. As can be observed from this graph, worms are able to survive on average for large periods of time (above two years), with an

[5] Data publicly available at the Web site http://www.messagelabs.com.

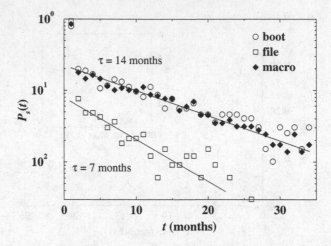

Fig. 9.3 Surviving probability for the three main strains of computer viruses. After a sharp initial drop, it is clear the presence of an exponential decay, with an associated characteristic time τ that depends on the given strain.

average prevalence well below 10^{-4}. As a single case study, Figure 9.2(b) reports the time evolution of the incidence of a typical worm, the *vbs/loveletter* virus, a variant of the *I-love-you* bug, starting from the month of its first observation. After a very fast increase the incidence decays until it reaches a long-lasting low prevalence state, with an average density of infected computers $\rho \sim 5 \times 10^{-6}$.

Associated with the long clean-up or endemic state, it is reasonable to expect a long average lifetime of computer viruses. In order to highlight this feature, it is possible to focus on the *surviving probability* of homogeneous groups of viruses, classified according to their infection mechanism (the strain) (Pastor-Satorras and Vespignani, 2001b). In these measurements one considers the total number of viruses of a given strain that are born and died within a given observation window. The surviving probability $P_s(t)$ of the strain is defined as the fraction of viruses still alive at time t after their birth. Figure 9.3 reproduces the survival probability estimated from prevalence data provided by the *Virus Bulletin*[6] from February 1996 to March 2000, covering a time interval of 50 months. Data show that the survival probability exhibits over long time periods a clean exponential tail

$$P_s(t) \sim \exp(-t/\tau), \tag{9.1}$$

where τ represents the characteristic lifetime of the virus strain. The numerical fit of the data (Pastor-Satorras and Vespignani, 2001b) yields $\tau \simeq 14$ months for boot and macro viruses and $\tau \simeq 6 - 9$ months for file viruses. Strikingly, these

---

[6] Data publicly available at the Web site http://www.virusbtn.com.

time scales are incredibly long if compared with the spreading rate of computer viruses and the deployement of anti-virus software, both of the order of a few days. As we shall see in the following sections, the evidence is not easily reconciliable with the usual framework of epidemic modeling and calls for the inclusion of the detailed topological structure of the Internet in the description of computer virus spreading.

9.2 Epidemic modeling in population networks

The theoretical modeling of computer virus epidemics has been stablished in close analogy with the models developed for the statistical study of biological diseases evolution.[7] Since the statistical modeling of epidemics aims at a large-scale description of the outbreak evolution, the details of the infection mechanism within each individual in the population are necessarily neglected. In this coarse grained view, individuals can only exist in a discrete set of states, such as susceptible (or healthy), infected (capable to spread the disease), immune, dead (or removed), etc. The description of the spreading process is in terms of *individuals* and their *interactions*, which are represented in the structure of the contacts along which the epidemics can propagate. From this point of view, the population is described as a graph, in which the vertices represent the individuals and the edges are the connections along which the epidemics propagates.[8]

The simplest epidemiological model one can consider, is the susceptible-infected-susceptible (SIS) model (Anderson and May, 1992; Diekmann and Heesterbeek, 2000). The SIS model is mainly used as a paradigmatic model for the study of infectious diseases leading to an endemic state with a stationary and constant value for the prevalence of infected individuals, i.e. the degree to which the infection is widespread in the population. In the SIS model, individuals can only exist in two discrete states, namely, susceptible and infected. The disease transmission is described in an effective way. The probability that a susceptible vertex acquires the infection from any given neighbor in an infitesimal time interval dt is $\lambda \, dt$, where λ defines the virus *spreading rate*. At the same time, infected vertices are cured and become again susceptible with probability $\delta \, dt$. Without lack of generality, we set can set the recovery rate $\delta = 1$, since it only affects the definition of the time scale of the disease propagation. Individuals thus run stochastically through the cycle susceptible $\rightarrow$ infected $\rightarrow$ susceptible, hence the name of the model. The SIS model does not take into account the possibility of the removal of individuals due to death or acquired immunization, which would

---

[7] For a review see Kephart and White (1991) and Kephart *et al.* (1993), and references therein.
[8] The graph defined by the population network is usually considered as undirected, which means that the propagation contacts (the edges) are always bi-directional.

lead to the so-called susceptible-infected-removed (SIR) model (Diekmann and Heesterbeek, 2000; Anderson and May, 1992; Murray, 1993) (see Appendix A6). The SIR model, in fact, assumes that infected individuals disappear permanently from the network at a unitary rate. This amounts to considering a situation in which infected computers are switched off at infection time and never rejoin the network unless completely screened by antivirus software. This model thus appears particularly suited for the initial stage of the epidemic when large numbers of computers become infected and antivirus is deployed by a number of concerned users. However, in the long run the clean-up stage reaches a stationary steady state in which the SIS model better represents the overall endemic state. Indeed, this stage refers to a population of users that, even though they use antiviruses, do not become more alert with respect to viral infection once they have cleaned their computers, which can again become infected (Kephart and White, 1991). This is due to several reasons such as miscorrect or disabled antivirus scanning configurations that occur rather frequently among users (Aron *et al.*, 2002). The SIS model thus represents epidemic outbreaks in the large time scale limit of the clean-up stage, and in the following we will use it to illustrate the general epidemic framework, deferring the analysis of the SIR case to the Appendix A6.

The analytical study of the SIS model can be undertaken in terms of the dynamic evolution for the epidemic prevalence of the epidemic outbreak. For homogeneous networks, in which the degree fluctuations are very small, we can approach this problem by writing the dynamic reaction rate equation for $\rho(t)$, defined as the density of infected vertices present at time t. That consists in a mean-field description of the system, in which all vertices are considered as *equivalent* irrespective of their corresponding degree, an assumption that is coherent as long as the variations in k are not very strong and correlations among the state of vertices can be neglected (Marro and Dickman, 1999). In this case, the average density $\rho(t)$ is also equivalent to the probability that any given vertex is infected and the reaction equation for $\rho(t)$ can be written as

$$\frac{d\rho(t)}{dt} = -\rho(t) + \lambda \langle k \rangle \rho(t) \left[1 - \rho(t) \right]. \tag{9.2}$$

The first term on the right-hand side of Eq. (9.2) considers infected vertices spontaneously recovering at a unit rate. The second term represents the rate of new infected vertices generated in the network, i.e. the density of healthy vertices acquiring the infection. This is proportional to the infection spreading rate, λ, the density of susceptible vertices that might become infected, $1 - \rho(t)$, and the number of infected individuals in contact with any healthy vertex. This last factor assumes the *homogeneous mixing hypothesis* (Anderson and May, 1992) which asserts that the force of the infection (the per capita rate of acquisition of the disease for the susceptible individuals) is proportional to the average number of contacts with infected individuals, that is approximated as $k\rho(t)$,[9] i.e. the average number of infected neighbors. In the homogeneous networks we are considering here, the degree has only very small fluctuations, $\langle k^2 \rangle \sim \langle k \rangle^2$, and as a first approximation we have considered that each vertex has the same number of edges, $k \simeq \langle k \rangle$. This homogeneity hypothesis is thus equivalent to assuming that the rate of contacts between infectious and susceptibles is constant for the whole population, and independent of any possible source of heterogeneity present in the system. In Eq. (9.2) we have also ignored all higher-order corrections in $\rho(t)$ (i.e. terms of order $\rho(t)^3$), since we are interested in the onset of the infection close to the point $\rho(t) \simeq 0$.

After imposing the stationarity condition $d\rho(t)/dt = 0$, we obtain the equation, valid for the behavior of the system at $t \to \infty$

$$\rho\left[-1 + \lambda\langle k \rangle(1 - \rho)\right] = 0, \tag{9.3}$$

whose solution yields the steady state density ρ of infected vertices. This equation defines an *epidemic threshold*

$$\lambda_c = \frac{1}{\langle k \rangle}, \tag{9.4}$$

which yields two different prevalence regimes as a function of the spreading rate:

$$\rho(\lambda) = \begin{cases} 0 & \text{if } \lambda < \lambda_c \\ (\lambda - \lambda_c)/\lambda & \text{if } \lambda \geq \lambda_c \end{cases}. \tag{9.5}$$

The presence of a non-zero *epidemic threshold* λ_c (Anderson and May, 1992; Murray, 1993; Diekmann and Heesterbeek, 2000) is a central result of the model. If the value of λ is above the threshold, $\lambda \geq \lambda_c$, the infection spreads and becomes endemic. Below the threshold, $\lambda < \lambda_c$, the infection dies out exponentially fast. As for percolation in Section 6.3, this behavior defines a phase transition between two very different regimes. This transition, however, is dynamical and more

---

[9] Since each infected individual attempts to infect a connected suceptible vertex with probability $\lambda \, dt$, a susceptible vertex with n infected neighbors will have a total probability of getting infected given by $1 - (1 - \lambda \, dt)^n$. Neglecting fluctuations, each susceptible vertex with k connections will have on average $n = k\rho$ infected neighbors, yielding at the leading order in $\lambda \, dt \ll 1$ an infection acquisition probability $1 - (1 - \lambda \, dt)^{k\rho} \simeq \lambda k\rho \, dt$. This finally recovers the per capita acquisition rate $\lambda k\rho$. Sometimes alternative definitions of the SIS model refer to $\lambda \, dt$ as the probability of acquiring the infection if one or more neighbors, indistinctively, are infected. In this case the total acquisition probability is given by $\lambda \, dt[1 - (1 - \rho)^k]$, i.e. the spreading probability times the probability that at least one neighbor is infected. Also in this case, for $\lambda \, dt \ll 1$ and $\rho \ll 1$ an acquisition rate $\lambda k\rho$ is recovered at the leading order.

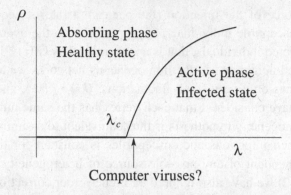

Fig. 9.4   Schematic phase diagram for the SIS model in homogeneous networks. The epidemic threshold λ_c separates an active or infected phase, with finite prevalence, from an absorbing or healthy phase, with null prevalence. Within this framework, the very small prevalence and long lifetimes observed in computer virus data are only compatible with a value of λ infinitesimally close to the epidemic threshold.

specifically the SIS model exhibits an *absorbing-state phase transition*.[10] Below the critical point the system relaxes in a state with null dynamics, the healthy phase. Above this point, a dynamical state characterized by a macroscopic number of infected individuals sets in, defining an infected phase. A qualitative picture of the phase diagram of this transition is depicted in Figure 9.4.

The concept of threshold is central also in issues related to the protection of populations by means of immunization programs. These correspond to vaccination policies, i.e. the delivery of anti-virus scanning software, aimed at the eradication of the epidemic. The simplest immunization procedure one can consider consists in the random introduction of immune individuals in the population (Anderson and May, 1992), in order to get a *uniform* immunization density. In this case, for a fixed spreading rate λ, the relevant control parameter is the density of immune vertices present in the network, the *immunity g*. At the mean-field level, the presence of a uniform immunity will have the effect of reducing the spreading rate λ by a factor $1 - g$; i.e. the probability of infecting a susceptible and non-immune vertex will be $\lambda(1 - g)[1 - \rho(t)]$. For homogeneous networks we can easily see that, for a constant λ, the stationary prevalence is given in this case by

$$\rho_g = \begin{cases} 0 & \text{if } g > g_c \\ \dfrac{g_c - g}{1 - g} & \text{if } g \le g_c, \end{cases} \qquad (9.6)$$

[10] It is possible to recognize that the SIS model is a generalization of the contact process model (Harris, 1974), widely studied in this context as the paradigmatic example of an absorbing-state phase transitions with a unique absorbing state (Marro and Dickman, 1999).

where g_c is the critical immunization value above which the density of infected individuals in the stationary state is null, and depends on λ as

$$g_c = 1 - \frac{\lambda_c}{\lambda}. \tag{9.7}$$

Thus, for a uniform immunization level larger than g_c, homogeneous networks are completely protected and no large epidemic outbreaks or endemic states are possible. The immunization threshold is very important in the prevention of epidemic outbreaks and in the clean-up stage in order to eradicate the virus. In practice, the aim of antivirus software deployment is to achieve a density of immunized individuals that pushes the population into the healthy region of the phase diagram.

Finally, it must be stressed that the concept of epidemic and immunization thresholds is a very general and key property of epidemic models. For instance, the SIR model too shows a definite epidemic threshold in homogeneous networks. In this model individuals are removed after having been infected and the asymptotic prevalence is necesseraly null. Yet, the SIR epidemic threshold separates a phase in which the epidemic outbreak involves only a negligible fraction of individuals from a phase in which a finite fraction of the population is infected (see Appendix A6). As well, clustered but homogeneous connectivity patterns such as regular lattices, meshes and the Watts–Strogatz network are not altering this scenario and just provide a different scaling behavior of the prevalence close to the threshold (Anderson and May, 1992; Marro and Dickman, 1999; Moore and Newman, 2000; Abramson and Kuperman, 2001).

9.3 Puzzling questions raised by computer virus data

When comparing the theoretical picture delivered by the SIS model on homogeneous networks with the behavior observed in real computer viruses, one is faced with some unexpected and paradoxical conclusions. First of all, the extremely long-lasting low prevalence shown by viruses in the clean-up stage is compatible with the phase diagram sketched in Figure 9.4 only in the very *unlikely* case that all surviving viruses are constructed such that their respective asymptotic spreading rate λ is tuned infinitesimally close to λ_c, above the epidemic threshold. At the same time, the characteristic life times observed in the analysis of the survival probability of the different virus strains are impressively long if compared with the time lapse before which anti-virus software is available on the market, usually days or weeks after the first incident report. Again such a long lifetime on the scale of the typical spread/recovery rates would suggest an effective spreading rate greater than the epidemic threshold, which is in contradiction with the always low prevalence levels of computer viruses except in the case of an unrealistic tuning of all viruses to the system's epidemic threshold. Finally, the deployment of antivirus alerts and software is nowadays considerable, with all major companies and providers constantly updating their antivirus software databases.[11] This fact, along with the timely availability of antivirus software, leads to a very large fraction of immunized individuals in a very short time scale compared with the duration of the epidemic outbreak. It is therefore very puzzling that despite these massive vaccination campaigns the threshold for virus eradication is never reached. Also in this case the only explantion within the usual framework would consist of a delicate interplay between the virus spreading rate and the immunization density that automatically tunes the effective spreading rate always very close to the critical point. Once more, this explanation is rather cumbersome and highly improbable. In summary, comparison with the known experimental data points out that the view obtained so far with the modeling of computer viruses is very instructive, but not easily reconciliable, even at a qualitative level, with the nature of the real digital epidemics phenomenon. The explanation of this discrepancy has been recognized as one of the most important open problems in computer virus epidemiology (White, 1998).

The key point that might provide a solution to the riddle posed by computer viruses resides in their transmission media (FTP, e-mail, etc.) (Kephart *et al.*, 1997). Viruses will spread preferentially to computers which are highly connected to the outer world and are thus proportionally exchanging more data and information. It is thus rather intuitive to consider the ubiquitus scale-free topology of technological networks as the effective one on which the spreading takes place. For instance, this is the case of *natural computer viruses* and *error propagation* processes which spread on the topology identified by routers and servers. Worms in turn spread on the social network defined by e-mail exchanges that in many instances exhibit scale-free properties (Ebel *et al.*, 2002). More recent worms and viruses take adantage of P2P systems and other virtual networks to diffuse (Staniford *et al.*, 2002), and also in these cases the heterogeneous topology is the one characterizing the individuals' connectivity patterns. A detailed discussion of the topology of social and virtual networks has been provided in Chapter 7, where the scale-free topology appears as a natural ingredient to be introduced in any modeling of physical processes occurring in these systems.

The conclusion from the above arguments is that computer viruses spread in highly heterogeneous networks, in which, even though the average degree is well defined, the degree fluctuations are unbounded; i.e. there is always a finite probability that a vertex has a number of neighbors much larger than the average value.

---

[11] For instance, most major business companies have set automatic weekly downloads of antivirus software for their employees.

These degree fluctuations are the key difference with respect to the epidemic models discussed on homogeneous networks, and their inclusion provides a different theoretical framework than appears to fit naturally the empirical evidence obtained from computer virus data.

9.4 Epidemics in scale-free networks

In order to take into account fully degree fluctuations in an analytical description of the SIS model, we have to relax the homogeneity assumption and allow for degree fluctuations by introducing the relative density $\rho_k(t)$ of infected vertices with given degree k; i.e. the probability that a vertex with k edges is infected. The dynamical reaction rate equations for any given degree class can thus be written as (Pastor-Satorras and Vespignani, 2001a, 2001b)

$$\frac{d\rho_k(t)}{dt} = -\rho_k(t) + \lambda k \left[1 - \rho_k(t)\right] \Theta_k[\{\rho_{k'}(t)\}], \qquad (9.8)$$

where also in this case we have considered a unitary recovery rate and neglected higher-order terms ($\rho_k(t) \ll 1$). The creation term considers the density $1 - \rho_k(t)$ of healthy vertices with k edges, that might get infected via a neighboring vertex. The rate of this last event is proportional to the infection rate λ, the actual number of connections k, and the average density of infected individuals connected at the end of each edge, i.e. the probability $\Theta_k[\{\rho_{k'}(t)\}]$ that an edge emanating from a node of degree k points to an infected individual. We make the assumption that Θ_k is a function only of the degree k and of the set of densities of infected vertices $\{\rho_{k'}(t)\}$ in each degree class.

9.4.1 Uncorrelated random networks

For the sake of simplicity, we will consider in the first place the solution of Eq. (9.8) where the underlying network is a generalized random graph (see Section 5.1.2) with no degree correlations, that is, when the probability that a vertex of degree k is connected to a vertex of degree k' is independent of the degree of the originating vertex k (see Appendix A4). In this case, the function Θ_k cannot depend on the variable k, and then $\Theta_k \equiv \Theta$. In the steady (endemic) state, ρ_k are functions of λ. Thus, the average density Θ becomes also an implicit function of the spreading rate, and by imposing the stationarity condition $d\rho_k(t)/dt = 0$, we obtain

$$\rho_k = \frac{k\lambda\Theta(\lambda)}{1 + k\lambda\Theta(\lambda)}. \qquad (9.9)$$

This set of equations shows that the higher a vertex degree, the higher the probability it will be in an infected state. This inhomogeneity cannot be neglected in the computation of $\Theta(\lambda)$. For the case of a random uncorrelated network, the calculation of $\Theta(\lambda)$ is straightforward, because the probability that any given edge points to a vertex with k' edges is equal to $k'P(k')/\sum_k kP(k)$. This probability yields an average density of infected vertices pointed to by any given edge that reads as

$$\Theta(\lambda) = \frac{1}{\langle k \rangle} \sum_k kP(k)\rho_k(\lambda), \tag{9.10}$$

where it has been considered that $\sum_k kP(k) = \langle k \rangle$. Since ρ_k in the stationary state is in its turn a function of $\Theta(\lambda)$, it is possible to use Eq. (9.9) to obtain the self-consistent equation

$$\Theta = \frac{1}{\langle k \rangle} \sum_k kP(k)\frac{\lambda k\Theta}{1 + \lambda k\Theta}, \tag{9.11}$$

whose solution allows the explicit calculation of $\Theta(\lambda)$ (Pastor-Satorras and Vespignani, 2001a, 2001b). Once this solution is obtained it is possible to get an explicit form for the densities ρ_k and finally to evaluate the average prevalence ρ as

$$\rho(\lambda) = \sum_k P(k)\rho_k(\lambda). \tag{9.12}$$

The self-consistent Eqs (9.9) and (9.10) can be approximately solved, in the limit of small Θ (close to the epidemic threshold), for scale-free distribution with general degree exponent γ. However, even without accessing the full solution of Eqs (9.9) and (9.10), the epidemic threshold can be explicitly calculated from Eq. (9.11) by simply observing that λ_c is the value of λ, above which it is possible to obtain a nonzero solution Θ^*. Using a geometrical argument, as in the case of the analysis of percolation theory in random graphs in Chapter 6, the solution of Eq. (9.11) follows from the intersection of the curves $y_1(\Theta) = \Theta$ and $y_2(\Theta) = (1/\langle k \rangle) \sum_k kP(k)\lambda k\Theta/(1 + \lambda k\Theta)$. The latter is a monotonously increasing and convex function of Θ between the limits $y_2(0) = 0$ and $y_2(1) = (1/\langle k \rangle) \sum_k kP(k)\lambda k/(1 + \lambda k) < 1$. If we must have a solution $\Theta^* \neq 0$, then the slope of $y_2(\Theta)$ at the point $\Theta = 0$ must be larger than or equal to 1 (see Figure 9.5). This condition implies that

$$\frac{d}{d\Theta}\left(\frac{1}{\langle k \rangle} \sum_k kP(k)\frac{\lambda k\Theta}{1 + \lambda k\Theta} \right)\Bigg|_{\Theta=0} \equiv \lambda\frac{\langle k^2 \rangle}{\langle k \rangle} \geq 1. \tag{9.13}$$

The value of λ yielding the equality in Eq. (9.13) defines the epidemic threshold

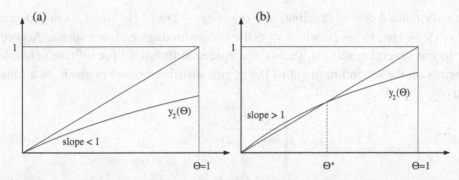

Fig. 9.5   Graphical solution of Eq. (9.11). (a) If the slope of the function $y_2(\Theta)$ at $\Theta = 0$ is smaller than 1, the only solution of the equation is $\Theta = 0$. (b) When the slope is larger than 1, a non-trivial solution $\Theta^* \neq 0$ can be found.

λ_c, that reads as

$$\lambda_c = \frac{\langle k \rangle}{\langle k^2 \rangle}. \tag{9.14}$$

This result implies that in scale-free networks with degree exponent $2 < \gamma \leq 3$, for which $\langle k^2 \rangle \to \infty$ in the limit of a network of infinite size, we have $\lambda_c = 0$, i.e. *a null epidemic threshold*. Also in this case, the parameter $\kappa = \langle k^2 \rangle / \langle k \rangle$ that defines the level of heterogeneity in the connectivity pattern is determining the properties of the physical processes occurring on the network. This is a very relevant result, which, analogously to those concerning the resilience to damage (Chapter 6), indicates that scale-free networks behave very differently, with respect to physical and dynamical processes, from homogeneous networks. In particular, the absence of any epidemic threshold makes scale-free networks a sort of ideal environment for the spreading of viruses, that also in the case of very weak spreading capabilities are able to pervade endemically the network. Moreover, the lack of an epidemic threshold in scale-free networks is not just a peculiar property of the SIS model. This scenario indeed holds also in other epidemic models and appears to be a general framework for epidemics in heterogeneous networks (Moreno, Pastor-Satorras, and Vespignani, 2002; May and Lloyd, 2001; Newman, 2002b). In Appendix A6, the calculation for the SIR model in scale-free networks is reported in detail as a convincing example of the generality of these results.

9.4.2 Prevalence behavior in scale-free networks

The previous analysis can be worked out in detail in random uncorrelated scale-free networks with an arbitrary degree exponent γ (Pastor-Satorras and Vespignani, 2001b). Consider a network which, in the continuous k approximation, has a normalized degree distribution $P(k) = (\gamma - 1)m^{\gamma-1}k^{-\gamma}$ and average degree $\langle k \rangle = (\gamma - 1)m/(\gamma - 2)$, where m is the minimum degree of any vertex. According to the general result Eq. (9.14), the epidemic threshold for infinite networks depends on the second moment of the degree distribution and is given, as a function of γ, by

$$
\lambda_c = \begin{cases} \dfrac{\gamma - 3}{m(\gamma - 2)} & \text{if } \gamma > 3 \\ 0 & \text{if } \gamma \leq 3 . \end{cases}
\tag{9.15}
$$

In this case it is also possible to calculate explicitly the average prevalence ρ as a function of λ and the specific scale-free exponent γ considered. By using the continuos k approximation, Eq. (9.11) yields the self-consistent relation

$$
\Theta(\lambda) = (\gamma - 2)m^{\gamma-2}\lambda\Theta(\lambda) \int_m^\infty \frac{k^{-\gamma+2}}{1 + k\lambda\Theta(\lambda)}\, dk.
\tag{9.16}
$$

Analogously, the expression for the density ρ, Eq. (9.12), reads as

$$
\rho(\lambda) = (\gamma - 1)m^{\gamma-1}\lambda\Theta(\lambda) \int_m^\infty \frac{k^{-\gamma+1}}{1 + k\lambda\Theta(\lambda)}\, dk.
\tag{9.17}
$$

In the limit $\rho \to 0$ (which obviously corresponds also to $\Theta \to 0$), we can perform an asymptotic expansion of the right-hand side of Eq. (9.17), which at the lowest order in λ yields the general form

$$
\rho(\lambda) \simeq \frac{\gamma - 1}{\gamma - 2}m\lambda\Theta(\lambda),
\tag{9.18}
$$

for all values of γ. The average prevalence is therefore a linear function of $\Theta(\lambda)$ whose explicit dependence on λ in turn depends on the particular value of γ considered through the self-consistent solution of Eq. (9.16). Depending on this solution the following cases can be considered:

(a) $2 < \gamma < 3$

In this case the leading order in λ of the solution of Eq. (9.16) yields

$$
\Theta(\lambda) \simeq (m\lambda)^{(\gamma-2)/(3-\gamma)} .
\tag{9.19}
$$

Combining this result with Eq. (9.18), we obtain

$$
\rho(\lambda) \sim \lambda^{1/(3-\gamma)}.
\tag{9.20}
$$

As expected, this relation does not show any epidemic threshold and gives non-zero prevalence for all values of λ. It is important to note that the exponent governing the behavior of the prevalence, $1/(3 - \gamma)$, is larger than 1. This implies that for small λ the prevalence is growing very slowly, i.e. there exists a wide region of the spreading rate in which $\rho \ll 1$.

(b) $\gamma = 3$

For this value of the degree exponent, logarithmic corrections dominate in the solution of Eq. (9.16), yielding

$$\Theta(\lambda) = \frac{e^{-1/m\lambda}}{\lambda m}(1 - e^{-1/m\lambda})^{-1}, \tag{9.21}$$

which gives an average prevalence at the lowest order in λ

$$\rho(\lambda) \sim 2e^{-1/m\lambda}. \tag{9.22}$$

Also in this case, the total absence of any epidemic threshold is recovered, and the prevalence approaches zero in a continuous and smooth way, exhibiting an exponentially small value for a wide range of spreading rates ($\lambda \ll 1$).

(c) $3 < \gamma < 4$

The non-zero solution for $\Theta(\lambda)$ yields

$$\rho(\lambda) \sim \left(\lambda - \frac{\gamma - 3}{m(\gamma - 2)}\right)^{1/(\gamma-3)} \tag{9.23}$$

That is, a power-law persistence behavior is observed. It is associated, however, with the presence of a non-zero threshold λ_c as given by Eq. (9.15). Since $1/(\gamma - 3) > 1$, the epidemic threshold is approached smoothly without any sign of the singular behavior associated with a critical point.

(d) $\gamma > 4$

The most relevant terms in the expansion of $\Theta(\lambda)$ yield now the behavior

$$\rho(\lambda) \sim \lambda - \frac{\gamma - 3}{m(\gamma - 2)}. \tag{9.24}$$

That is, we recover the usual epidemic framework obtained for homogeneous networks.

In summary, the outcome of the analysis presented here is that the SIS model in scale-free uncorrelated random networks with degree exponent $\gamma \leq 3$ exhibits the absence of any epidemic threshold or critical point, i.e. $\lambda_c = 0$. Only for $\gamma > 4$ do epidemics on scale-free networks have the same properties as on homogeneous networks. This result is consistent with the general picture already encountered for resilience to damage, and also in this case finds its origin in the presence of highly connected hubs. The latter, indeed, can be considered as sort of *super-spreaders*. Each time the epidemic reaches one of the hubs it has the possibility of infecting an enormous number of individuals ($k\lambda$, with $k \to \infty$) irrespective of the very low spreading rate. The epidemic thus receives a burst that keeps it alive until a new hub is hit. In homogeneous networks, and for $\gamma > 3$, the probability of finding a hub is negligible and the epidemic dies if the spreading rate is not large enough to infect a sufficient number of new individuals by using the connectivity of "typical" vertices. In scale-free networks, instead, there is a non-negligible probability that the infection will be transmitted from hub to hub, each time allowing the virus to infect a finite fraction of the population's individuals whatever the spreading rate is.

9.4.3 Finite-size effects

The actual Internet and the social or virtual networks considered here are composed of a finite number of elements, which is far from the thermodynamic limit. This finite population introduces a maximum degree k_c, depending on the system size N (see Appendix A5) or a finite connectivity capacity, which has the effect of restoring a bound to the degree fluctuations. The presence of the cut-off translates, through the general expression Eq. (9.14), into an effective non-zero epidemic threshold due to *finite size effects*, as usually observed in non-equilibrium phase transitions (Pastor-Satorras and Vespignani, 2002; May and Lloyd, 2001; Marro and Dickman, 1999). This positive epidemic threshold, however, is not an *intrinsic* property as in homogeneous systems, but an artifact of the limited system size that vanishes when increasing the network size or the degree cut-off.

Focusing in the SIS model in uncorrelated random networks with a scale-free degree distribution of the form $P(k) \simeq k^{-\gamma} \exp(-k/k_c)$, we can compute the effective non-zero epidemic threshold $\lambda_c(k_c)$ within the continuous k approximation (Pastor-Satorras and Vespignani, 2002)

$$\lambda_c(k_c) = \frac{\langle k \rangle_{k_c}}{\langle k^2 \rangle_{k_c}} = \frac{\int_m^\infty k^{-\gamma+1} \exp(-k/k_c)}{\int_m^\infty k^{-\gamma+2} \exp(-k/k_c)} \equiv \frac{\Gamma(2-\gamma, m/k_c)}{\Gamma(3-\gamma, m/k_c)}, \quad (9.25)$$

where m is the minimum degree of the network, and $\Gamma(x, y)$ is the incomplete Gamma function (Abramowitz and Stegun, 1972). For large k_c we can perform a

Taylor expansion and retain only the leading term, obtaining for any $2 < \gamma < 3$

$$\lambda_c(k_c) \simeq \left(\frac{k_c}{m}\right)^{\gamma - 3}. \tag{9.26}$$

The limit $\gamma \to 3$ corresponds to a logarithmic divergence, yielding instead the leading order

$$\lambda_c(k_c) \simeq \frac{1}{m \ln(k_c/m)}. \tag{9.27}$$

It is interesting to compare the intrinsic epidemic threshold obtained in homogeneous networks with negligible degree fluctuations, $\lambda_c^H = \langle k \rangle^{-1}$, with the non-zero effective threshold of bounded scale-free distributions. Figure 9.6 represents the ratio $\lambda_c(k_c)/\lambda_c^H$ as a function of k_c/m, for different values of the degree exponent γ. We can observe that, even for relatively small cut-offs ($k_c/m \sim 10^2 - 10^3$), for a reasonable value $\gamma \approx 2.5$ the effective epidemic threshold of finite scale-free networks is smaller by close to an order of magnitude than the intrinsic threshold corresponding to a homogeneous network with the same average degree. This fact implies that the extreme weakness of scale-free networks to epidemic agents is present even in finite-size or degree-bounded networks. The use of the homogeneity assumption would lead in scale-free networks to a serious over-estimate of the epidemic threshold, even for relatively small networks.

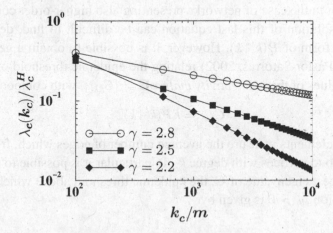

Fig. 9.6   Ratio between the effective epidemic threshold $\lambda_c(k_c)$ in bounded scale-free networks with a soft exponential cut-off, and the intrinsic epidemic threshold λ_c^H for homogeneous networks with the same average degree, for different values of γ.

9.4.4 Correlated random networks

For a general network in which the degrees of the different vertices are correlated (Appendix A4), the dynamical equation considered so far might result in a too strong approximation, since we are not considering the effect of the degree k into the expression for Θ_k. A more refined scheme allows the inclusion of correlations defined by the conditional probability $P(k' \mid k)$ that a vertex of degree k is connected to a vertex of degree k'. The inclusion of correlations yields the correct factor Θ_k as

$$\Theta_k[\{\rho_{k'}(t)\}] = \sum_{k'} P(k' \mid k)\rho_{k'}(t), \tag{9.28}$$

that is, the probability that an edge in a vertex of degree k is pointing to an infected individual is proportional to the probability that any edge points to a vertex of degree k', times the probability that this vertex is infected, $\rho_{k'}(t)$, averaged over all the connections of the original vertex. From Eqs. (9.8) and (9.28), the mean-field equations describing the SIS epidemic model on correlated random networks can be written as

$$\frac{\mathrm{d}\rho_k(t)}{\mathrm{d}t} = -\rho_k(t) + \lambda k \left[1 - \rho_k(t)\right] \sum_{k'} P(k' \mid k)\rho_{k'}(t). \tag{9.29}$$

The above set of equations can be considered as exact for *Markovian random networks* (Boguñá and Pastor-Satorras, 2002), which are entirely defined by their degree distribution $P(k)$ and the conditional probability $P(k' \mid k)$, while it is a better approximation in the case of networks presenting also higher-order correlations.

The exact solution of this last equation can be difficult to find, depending on the particular form of $P(k' \mid k)$. However, it is possible to obtain a general result (Boguñá and Pastor-Satorras, 2002) relating the epidemic threshold of Eq. (9.29) to the eigenvalues of the *connectivity matrix* $\mathbf{C} = \{C_{kk'}\}$, with components

$$C_{kk'} = k P(k' \mid k). \tag{9.30}$$

These matrix elements measure the average number of edges which, from a vertex of degree k, go to vertices with degree k'. In particular it is possible to show that if Λ_m is the largest eigenvalue of $\mathbf{C}$, the epidemic threshold above which Eq. (9.29) allows a solution $\rho_k > 0$ is given by

$$\lambda_c = \frac{1}{\Lambda_m}. \tag{9.31}$$

It is instructive to see how this general formalism recovers the result in Eq. (9.14), implicitly obtained for random uncorrelated networks. For any random network, in which there are no correlations among the degrees of the vertices, we have that the connectivity matrix is given by $C_{kk'}^{nc} = kP(k' \mid k) \equiv kk'P(k')/\langle k \rangle$ (see Eq. (A4.8)). Now it is easy to see that the matrix $\{C_{k'k}^{nc}\}$ has a unique eigenvalue $\Lambda_m^{nc} = \langle k^2 \rangle/\langle k \rangle$, corresponding to the eigenvector $v_k^{nc} = k$, from where we recover the now well-established result Eq. (9.14).

While explicit solutions of the Eq. (9.29) cannot generally be obtained, it is very important to study to what extent the presence of degree correlations can alter the general picture stating that scale free networks do not have an epidemic threshold. In correlated networks the presence or lack of an epidemic threshold is directly related to the largest eigenvalue Λ_m of the connectivity matrix, which has been shown to diverge for all scale-free *unstructured networks*[12] with infinite fluctuations, for any kind of two-point correlations (Boguñá, Pastor-Satorras, and Vespignani, 2003a). Thus it is possible to state that a scale-free degree distribution with exponent $2 < \gamma \leq 3$ is a sufficient condition for a vanishing epidemic threshold in the thermodynamic limit in the case of correlated networks, confirming the general scenario in which scale-free networks appear to be distinctively weak with respect to error and virus transmission.

9.5 Numerical simulation of epidemics in network models

In order to have a test of the theoretical results and a numerical example of the epidemic framework in scale-free networks, the most convenient solution consists in performing numerical simulations of epidemic outbreaks in network models. For this purpose, it is possible to develop an agent-based modeling strategy in which each individual vertex is tracked as being either susceptible or infected. At each time step the updating dynamics as defined in the SIS model is applied to each vertex, depending on the state of all vertices. In this case, discrete time steps are conveniently used with parallel dynamics in which at each time step all healthy vertices in contact with an infected individual will become infected with probability λ. At the same time, all previously infected vertices switch with probability 1 to the healthy state at the end of the time step. The system is then let free to evolve following the stochastic microscopic dynamics, and during each time step it is possible to record average quantities such as the prevalence $\rho(t)$. In addition, given the stochastic nature of the dynamics, the experiment can be repeated with different stochastic realizations with the possibility of measuring quantities such as the average lifetime of outbreaks or their relaxation time scale to the endemic state. Different initial conditions can be choosen as well as the graph

---

[12] The present result is only valid for networks with no internal structure, in which all the vertices with the same degree are statistically equivalent. It does not apply for regular lattices or structured networks (Klemm and Egüíluz, 2002b; Moreno and Vázquez, 2003), in which a spatial or class ordering constrains the connections among vertices.

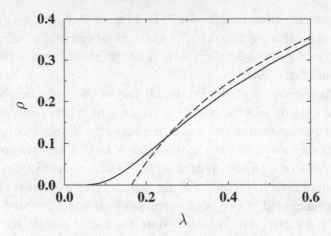

Fig. 9.7   Total prevalence ρ for the SIS model in a Barabási–Albert network (full line) as a function of the spreading rate λ, compared with the theoretical prediction for a homogeneous network (dashed line).

defining the connectivity pattern of the population. Interestingly, this approach is equivalent to the real evolution of an SIS epidemic outbreak in the generated network and can be used to validate the theoretical results obtained with the mean-field approximation.

Numerical simulations of the SIS model performed on a Barabási–Albert network with degree distribution $P(k) \sim k^{-3}$ confirm the analytical picture extracted from the mean-field analysis. Figure 9.7 shows the total prevalence ρ in the steady state as a function of the spreading rate λ (Pastor-Satorras and Vespignani, 2001a). As we can observe, it approaches zero in a continuous and smooth way, compatible with the presence of a vanishing epidemic threshold (see for comparison the behavior expected for a homogeneous network, also drawn in Figure 9.7). More-precisely, it is possible to focus on the region of small spreading rate $\lambda \ll 1$ in order to inspect the analytic form of the prevalence behavior. Figure 9.8 represents ρ in a semilogarithmic plot as a function of $1/\lambda$, which shows that $\rho(\lambda) \sim \exp(-C/\lambda)$, where C is a constant independent of the size N of the network, in very good agreement with the theoretical prediction of Eq (9.22).

The surviving probability $P_s(t)$ for a fixed value of λ and different network sizes N is represented in Figure 9.9. In this case, we recover exponential behavior in time, that has its origin in the finite size of the network. In fact, for any finite system, the epidemic will eventually die out because there is a finite probability that all individuals cure the infection at the same time. This probability is decreasing with the system size, and the lifetime is infinite only in the thermodynamic limit $N \to \infty$. However, the lifetime becomes virtually infinite (the metastable state has a lifetime too long for our observation window) for large enough sizes that depend

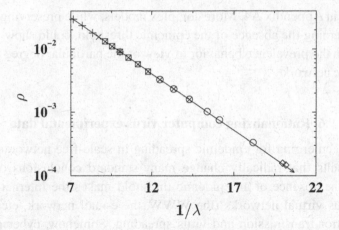

Fig. 9.8 Prevalence ρ as a function of $1/\lambda$ for Barabási–Albert networks of different sizes: $N = 10^5$ (+), $N = 5 \times 10^5$ ($\square$), $N = 10^6$ ($\times$), $N = 5 \times 10^6$ ($\circ$), and $N = 8.5 \times 10^6$ ($\diamond$). The linear behavior on the semi-logarithmic scale proves the stretched exponential behavior predicted for the prevalence. The full line is a fit to the form $\rho \sim \exp(-C/\lambda)$.

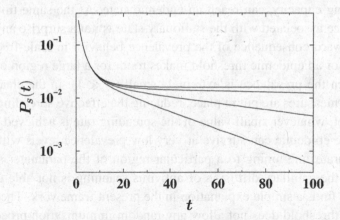

Fig. 9.9 Surviving probability $P_s(t)$ as a function of time in supercritical spreading experiments in the Barabási–Albert network. Spreading rate $\lambda = 0.065$. Network sizes ranging from $N = 6.25 \times 10^3$ to $N = 5 \times 10^5$ (bottom to top).

upon the spreading rate λ. This is a well-known feature of the survival probability in finite size absorbing-state systems poised above the critical point (Marro and Dickman, 1999). In our case, this picture is confirmed by numerical simulations that show that the average lifetime of the survival probability is increasing with the network size for all the values of λ. As a final comment, it must be noted that the Barabási–Albert network is quantitatively described by Eq. (9.22), which refers to the uncorrelated approximation because of the dimness of its degree correlations, as discussed in Appendix A4. More complex models, while preserving the general picture concerning the absence of the epidemic threshold, could show quantitative differences in the prevalence behavior in view of the particular degree correlations present in the network.

9.6 Rationalizing computer virus experimental data

The scenario emerging for epidemic spreading in scale-free networks shows unexpected results that radically change many standard conclusions on epidemic spreading. The absence of an epidemic threshold makes the Internet and highly heterogeneous virtual networks (the WWW, the e-mail network, etc.) very vulnerable to error transmission and virus spreading. Somehow, cyber plagues can proliferate on these networks whatever spreading rates they may have.

While this scenario is raising many concerns, it also provides a simple and reasonable rationalization of the features observed in experimental data from computer viruses. The very long-lasting clean-up stage common to many epidemic outbreaks is naturally accounted for by the fact that all viruses, independently of their spreading capacity, can reach an endemic state. At the same time, the very low prevalence associated with the stationary state is not a surprise anymore. It is a straightforward consequence of the prevalence behavior in scale-free networks. The absence of an epidemic threshold makes room for a large region of spreading rates in which the prevalence is extremely small ($\rho \ll 1$). In the clean-up stage, virus countermeasures are put in place, reducing the effective spreading rate of the epidemic. Yet, whatever small value of the spreading rate is achieved by security measures, the epidemic can survive at very low prevalence levels without invoking any accurate fine tuning to a particular region of the parameter space. Also the fact that the capillary diffusion of antivirus scanning is not able to eradicate the epidemic finds a simple explanation in the present framework. The absence of an epidemic threshold does not allow any random immunization procedure to be effective since any healthy region of the phase space in which the system can be poised does not exist. We shall look at this point in detail in the following sections.

Finally, it must be said that simple models such as the SIS and the SIR, while very instructive, cannot be considered realistic, and many more ingredients should be added to the representation of real epidemics. Rules defining the temporal patterns of networks such as the formation of new connections, the actual time during which connections exist, and other heterogeneities should be included in the models. Details of each particular virus can be relevant to some predictions and the users' behavior should be more carefully incorporated[13] in the description of

---

[13] This is particularly true in the case of active worms that guess susceptible computers also by specific algorithms not directly related to the network connectivity (Staniford *et al.*, 2002).

epidemic outbreaks. Nevertheless, the conceptual understanding of the peculiarities of virus spreading in scale-free networks can be considered as a first and basic step in fighting and forecasting epidemics in the cyber world.

9.7 Immunization of scale-free networks

The weakness to epidemic attacks of scale-free networks is presenting us an extremely worrying scenario. The conceptual understanding of the mechanisms and causes for this weakness, though, allows us to develop new defensive strategies that take advantage of the scale-free topology. Thus, while random immunization strategies are utterly inefficient, it is possible to devise targeted immunization schemes which are extremely effective.

9.7.1 Uniform immunization

In scale-free networks the introduction of a random immunization is able to locally depress the infection's prevalence, but it does so too slowly, being impossible to find any critical fraction of immunized individuals that ensures the infections eradication. A simple argument for the inadequacy of random immunization strategies is that they are giving the same importance to very connected vertices (with the largest infection potential) and to vertices with a very small degree. Due to the large fluctuations in the degree, heavily connected vertices, which are statistically very significant, can overcome the effect of the uniform immunization and maintain the endemic state.

In mathematical terms, the introduction of a density g of immune individuals chosen at random is equivalent to just rescaling the effective spreading rate as $\lambda \rightarrow \lambda(1 - g)$, i.e. the rate at which new infected individuals appear is depressed by a factor proportional to the probability that they are not immunized. However, the absence of an epidemic threshold ($\lambda_c = 0$) in the thermodynamic limit implies that whatever rescaling of the spreading rate does not bring the epidemic in the healthy region but for the case $g = 1$. Indeed, the immunization threshold g_c is obtained when the rescaled spreading rate is set equal to the epidemic threshold. For instance, using Eq. (9.14) for uncorrelated networks we obtain

$$\lambda(1 - g_c) = \frac{\langle k \rangle}{\langle k^2 \rangle}. \tag{9.32}$$

In scale-free networks with $\langle k^2 \rangle \rightarrow \infty$ only a complete immunization of the network (i.e. $g_c = 1$) ensures an infection-free stationary state in the thermodynamic limit. The fact that uniform immunization strategies are less effective has been noted in the biological context in several cases of spatial heterogeneity (Anderson and May, 1992). In scale-free networks, however, we face a limiting case due to the extremely high (virtually infinite) heterogeneity in the connectivity properties.

Noticeably, data from real computer virus epidemics provide support for this picture. Despite the deployment of antivirus software is timely and capillary, viruses' lifetimes are extremely long; in other words, very high levels of uniform immunization are not able to eradicate the epidemic. These empirical findings are, however, in good agreement with the picture obtained for the immunization of scale-free networks. In fact, antivirus deployment is not eradicating epidemics on the global scale, since it is like a random immunization process where file scanning and antivirus updating are overall left to the good will of users and system managers. Needless to say, from the point of view of the single user, antiviruses are extremely important, being the only way to ensure local protection for the computer.

9.7.2 Targeted immunization

Scale-free networks hinder the efficiency of naïve uniform immunization strategies. However, we can take advantage of their heterogeneity by devising immunization procedures that take into account the inherent hierarchy in the degree distribution. In fact, we know that scale-free networks posses a noticeable resilience to *random* connection failures (Chapter 6), which implies that the network can resist a high level of accidental damage without losing its global connectivity properties; i.e. the possibility to find a connected path between almost any two vertices in the system. At the same time, scale-free networks are strongly affected by *targeted* damage; if a few of the most connected vertices are removed, the network suffers a dramatic reduction in its ability to carry information. Applying this argument to the case of epidemic spreading, we can devise a *targeted* immunization scheme in which we progressively make immune the most highly connected vertices, i.e. the ones more likely to spread the disease. While this strategy is the simplest solution to the optimal immunization problem in heterogeneous populations (Anderson and May, 1992), its efficiency is comparable with the uniform strategies in homogeneous networks with finite degree variance. In scale-free networks, on the contrary, it produces a striking increase in the network's tolerance to infections at the price of a tiny fraction of immune individuals.

An approximate calculation of the immunization threshold in the case of a random scale-free network (Pastor-Satorras and Vespignani, 2001c) can be pursued along the lines of the analysis of the intentional attack of complex networks (see Section 6.6.1). Let us consider the situation in which a fraction g of the individuals with the highest degree have been successfully immunized. This corresponds, in the limit of a large network, to the introduction of an upper cut-off $k_c(g)$ – which is obviously an implicit function of the immunization g – such that all vertices with degree $k > k_c(g)$ are immune. At the same time, the infective agent cannot propagate along all the edges emanating from immune vertices, which translates into a probability $r(g)$ of deleting any individual contacts in the network. The elimination of edges and vertices for spreading purposes yields a new connectivity pattern whose degree distribution and the relative moments $\langle k \rangle_g$ and $\langle k^2 \rangle_g$ can be computed as a function of the density of immunized individuals.[14] The protection of the network will be achieved when the effective network on which the epidemic spreads satisfies the inequality $\langle k \rangle_g / \langle k^2 \rangle_g \geq \lambda$, yielding the implicit equation for the immunization threshold

$$\frac{\langle k \rangle_{g_c}}{\langle k^2 \rangle_{g_c}} = \lambda. \tag{9.33}$$

The epidemic threshold is naturally an implicit function $g_c(\lambda)$ and its analytic form will depend on the original degree distribution of the network.

In order to assess the efficiency of the targeted immunization scheme it is possible to perform the explicit calculation for an uncorrelated network with degree exponent $\gamma = 3$ (Pastor-Satorras and Vespignani, 2001c). In this case the leading order solution for the immunization threshold, in the case of targeted immunization, reads as

$$g_c \simeq \exp(-2/m\lambda). \tag{9.34}$$

This clearly indicates that the targeted immunization program is extremely convenient, with a critical immunization threshold that is exponentially small over a wide range of spreading rates λ. This theoretical prediction can be tested by performing direct numerical simulations of the SIS model on Barabási–Albert networks in the presence of targeted immunization. In Figure 9.10 the results of targeted immunization are compared with simulations made with a uniform immunization (Pastor-Satorras and Vespignani, 2001c). The plots show the reduced prevalence ρ_g/ρ_0, where ρ_g is the prevalence in the network with immunization density g and ρ_0 is the prevalence in the non-immunized network, at a fixed spreading rate $\lambda = 0.25$. This plot indicates that, for uniform immunization, the prevalence decays very slowly when increasing g, and will be effectively null only for $g \to 1$, as predicted by Eq. (9.32).[15] However, for the targeted immunization scheme, the prevalence shows a very sharp drop and exhibits the onset of a sharp immunization threshold above which the system is infection free. A linear regression from the largest values of g yields an approximate immunization threshold $g_c \simeq 0.06$,

---

[14] See the analogous calculation for the targeted removal of vertices in Section 6.6.1.
[15] The threshold is not exactly one because of the usual finite size effect present also in the simulations which are performed on networks of size $N = 10^7$.

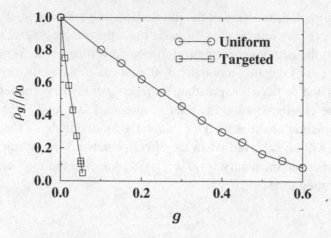

Fig. 9.10   Reduced prevalence ρ_g/ρ_0 from numerical simulations of the SIS model in the Barabási–Albert network with uniform and targeted immunization, at a fixed spreading rate $\lambda = 0.25$. A linear extrapolation from the largest values of g yields an estimate of the threshold $g_c \simeq 0.06$ for targeted immunization.

that definitely proves that scale-free networks are very sensitive to the targeted immunization of a very small fraction of the most connected vertices.

9.8 Protecting the Internet

The protection of the Internet from computer viruses and bugs is a major challenge due to the scale-free nature of the network over which the epidemics spread. On a global level, uniform immunization policies are not satisfactory and the spreading of errors or infective agents on scale-free networks can be contrasted only by a careful choice of immunization procedure. In particular, these procedures should rely on the identification of the most connected individuals. Fortunately, the protection of just a tiny fraction of these individuals raises dramatically the tolerance to infections of the whole population.

As a practical example of the effect of targeted immunization, it is possible to perform numerical experiments on real Internet maps. In Figure 9.11 the behavior of the normalized prevalence ρ_g/ρ_0 in simulation of the SIS model on the Internet AS level graph with a progressively increasing immunization density g is reported. The numerical experiment is performed by supposing an SIS dynamics for a BGP level propagating error, transmitted at constant spreading rate $\lambda = 0.25$ and recovered with unitary rate by external interventions that set the problem time scale. In addition, a density g of ASs is considered immune to the error propagation by using uniform and targeted strategies, alternatively. The figure clearly shows that the results are completely analogous to those obtained in simulations on network

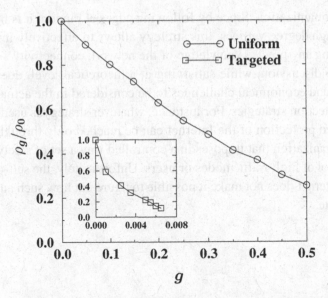

Fig. 9.11 Normalized prevalence ρ_g/ρ_0 from numerical simulations of the SIS model in the AS Internet map with uniform (main figure) and targeted (inset) immunization at a constant spreading rate $\lambda = 0.25$ and for incresing density g of immunized individuals.

models. Random immunization does not allow any drastic reduction of the prevalence of error affected ASs. For instance, the immunization of 25% of the vertices reduces by less than a factor half the average prevalence with respect to the case with no immune vertices. In the case of targeted strategies, however, the same level of protection is achieved by just immunizing less than 0.2% of the total population.

While the targeted strategy we have reported previously is very effective, it suffers from a practical drawback in its real world application. Its implementation requires *complete* knowledge of the network structure in order to identify and immunize the most connected vertices. For this reason, several strategies overcoming this problem have been proposed, mainly relying just on local, rather than global, knowledge of the network. In this context, Dezsö and Barabási (2002) propose a level of safety and protection policy, which is proportional to the importance of the vertex measured as a function of its local degree. This implies that high degree vertices are infected with a rate inversely proportional to their degree, or more in general as $k^{-\alpha}$. At the theoretical level it is possible to show that any $\alpha > 0$ reintroduces a finite threshold that, in the continuous k approximation, is estimated to be $\lambda_c(\alpha) = \alpha m^{\alpha-1}$. Another ingenious immunization strategy was proposed by Cohen, ben-Avraham, and Havlin (2002a), levering on a local exploration mechanism. In this scheme, a fraction g of vertices are selected and each one is asked to point to one of its neighbors. The neighbors, rather than the selected vertices, are chosen for immunization. Since by following edges at random it is more probable to point to high degree vertices, this strategy allows to effectively immunize hubs without having any precise knowledge of the network connectivity.

The above discussion, while satisfying at a theoretical level, does not address the political and economical challenges to be considered in the actual deployment of global protection strategies. For instance, whatever strategy is used, it is obvious that optimized protection of the Internet can be reached only through a global supervising organization that imposes in a controlled way a set of safety measures to a selected pool of high-traffic nodes or users. Unfortunately, the self-organized nature of the Internet does not make it possible to figure out how such an organization should operate.

10

Beyond the Internet's skeleton: traffic and global performance

In previous chapters we have presented an X-ray picture of the Internet's topological skeleton and the consequences that the revealed structure has on several properties of this network. There is, however, another dimension to add to the Internet picture that refers to the quantitative characterization of the actual flow of data and transmission capacity of the various elements forming the Internet. Physical links have a large heterogeneity in their data transmission capacity. Also, the traffic load is highly variable on different connections and at different times; that is, a full characterization of the network cannot prescind from the knowledge of the interplay between the topological structure, the detailed physical properties of each element, and the traffic generated by users.

The measurement and modeling of link traffic and performance[1] in LANs or WANs has been a major area of activity over the recent years. These studies pointed out the failure of the Poisson traffic model, providing empirical evidence for traffic *self-similarity*. The modeling of these properties finds a natural framework in the vast mathematical theory of self-similar stochastic processes, and intense research activity has been focused both on the understanding of the origin of the scale-free nature of traffic and its implications at the performance level. Measurements of link traffic imply the accumulation of very long time series of traffic loads and usually refer to a specific link or very small networks. Mathematical modeling and study, as well, are generally focused on small LANs or simple client–server architectures. This work is fundamental to understand the basic characteristics of traffic features, but by its very nature is difficult to extend at a large-scale statistical level.

Indeed, the large-scale statistical characterization of traffic and performance of the Internet is a task that is still at an early stage. As we have seen in previous

---

[1] With the term performance one generally refers to the many network characteristics that provide a measurement of network reliability and efficiency: end-to-end delay times, packet loss, latency, etc.

chapters, the simple characterization of the Internet's connectivity structure is already posing a particularly difficult challenge at both the measurement and modeling levels. The measurements of traffic paths, packet loss, and delay times for the Internet at large add new problems, ranging from data storage capacity to the eventual intrusive character of the measurements, that could generate a large amount of extra traffic. The sparseness of data on the load and performance of the Internet at the large-scale level is naturally reverberating on modeling efforts, where the lack of clear empirical evidence leaves room for very general and sometimes vague initial assumptions on the formulation of the model. At the same time the successful validation of any modeling framework is often not possible because of the lack of accurate data sets.

In this penultimate chapter we want to provide an overview of these issues. First we intend to present a brief introduction to the self-similar traffic and performance measurement and modeling. This is a vast area of activity and a detailed account goes beyond the scope of the present book, although it might be relevant to the understanding of the interplay between local traffic properties and the network topology and performance at the large-scale level. Afterwards, we shall discuss recent results on the measurements and data analysis of large-scale Internet performance and traffic data. The research effort is still scattered and preliminary, and a comprehensive presentation of the various results is not yet possible. Nevertheless, being at the end of the book, we want to convey to the reader the outlook for some activities and results that we consider of possible impact in the very near future.

10.1 The Internet traffic: a local view

The behavior of Internet traffic has been analyzed since the early days of computer networks.[2] These studies focus on the time series of the traffic load, measured as the number of IP packets, flowing through a given Internet link. Time series can be collected on different time scales, producing signals such as the one shown in Figure 10.1, in which each observation point represents the aggregate traffic $Y(i)$ recorded on the link in a time interval i of length τ. The time resolution τ can be opportunely chosen, resulting in different magnitudes of the signal Y, and the time series can be recorded for a total time spanning from a few seconds to weeks. Together with the number of packets, other quantities related to network performance can also be measured. For instance, the packet-loss rate, defined as the number of packets that are discarded or lost by the network after excessive

---

[2] This kind of analysis has been supported by the timely availability of packet traffic tools such as tcpdump, developed by V. Jacobson, C. Leres, and S. McCanne in 1989.

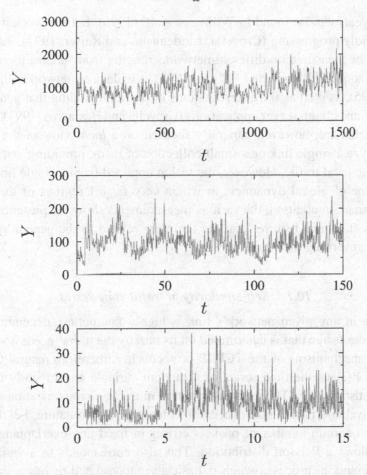

Fig. 10.1 The figure reports three snapshots at different resolution scales of a self-similar signal representing an hour's worth of TCP wide-area traffic between Digital Equipment Corporation and the rest of the world. Each plot represents the number of packets recorded on a link during time intervals τ that scales with a factor of ten larger in each plot. The time scale is correspondingly larger by a factor of ten in each plot. The y axis is corresponding rescaled since it represents the aggregated traffic on increasing time intervals. It is easy to recognize a large level of similarity in the three plots despite that they have scale variations spanning three orders of magnitude. Data collected in 1995 by Jeff Mogul of Digital's Western Research Lab (WRL) and publicly available on The Internet Traffic Archive, http://ita.ee.lbl.gov/html/traces.html. A detailed analysis of these traces has been provided by Paxson and Floyd (1995).

delays in router queues or for other reasons, as well the delay times in queues can provide an indication of the level of congestion on the network.

The mathematical modeling of traffic signals and performance in networks is a vast field of research. An impressive bibliographical guide to the literature in the early years can be found in Willinger *et al.* (1996), but the work in the field is still rapidly progressing (Crovella, Lindemann, and Reiser, 1997). Traffic time series can be measured on different networks ranging from small Ethernet LANs (Leland, Taqqu, Willinger, and Wilson, 1994) to wide area networks (Paxson and Floyd, 1995), as well as for different kinds of traffic, including that generated by the WWW and client–server applications (Crovella and Bestavros, 1997). All these measurements are, however, generally focused on a *local* view of the network. They analyze a single link or a small collection of them, providing just the properties of the local traffic. However, the vision obtained from a single link is often the outcome of global dynamics, in which very far off regions of the network also cooperate. In addition, the various measurements show the presence of some ubiquitous statistical features that are therefore related to the general dynamical principles governing the whole network.

10.1.1 Self-similarity of traffic time series

The traffic in any given network's link is highly fluctuating, depending on the instantaneous usage that is determined on its turn by the users' needs and the self-regulating mechanisms of the TCP/IP protocol. It is therefore natural to look at the signal $Y(i)$ as the time series of a random variable and to study its behavior in a statistical way. The usual assumption in traffic characterization is that the packet arrival and packet size distributions have a Poisson nature, i.e. the probability that a certain number of packets arrives in fixed non-overlapping time intervals follows a Poisson distribution. This also corresponds to a variable $Y(i)$ defining a random process, which is basically uncorrelated or has a short range autocorrelation function. Thus, if we consider the stationary stochastic process $Y = \{Y(i) : i = 1, 2, 3 \ldots\}$ with constant mean $\mu = \langle Y(i) \rangle$ and finite variance $\sigma^2 = \langle (Y(i) - \mu)^2 \rangle$, it is defined as short ranged if its autocorrelation function $C_Y(t) = \langle (Y(i) - \mu)(Y(i + t) - \mu) \rangle$ decays exponentially fast at large t. In such a kind of process, the larger is the time scale at which we observe the signal (or coarser the aggregation interval τ), the smaller will be the level of fluctuation around the average value of the process. In particular, if one considers the aggregated stochastic process $Y^m = \{Y^m(i) : i = 1, 2, 3...\}$ that indicates the new time series obtained by averaging the original series Y over non-overlapping blocks of size m, then the variance σ_m^2 of the process Y^m scales regularly as m^{-1}. In this case larger resolution plots of the time series are clearly smoothing-out the level of fluctuations with respect to the average value of the signal.

The Internet traffic, however, shows very different behavior. In particular, the signal burstiness is extremely persistent at very large resolution or large aggregation intervals. In more precise terms, the measured Internet traffic is self-similar with respect to time scaling. In Figure 10.1 we report an example of this property by showing a self-similar time series seen at different resolution time scales. A self-similar signal can be modeled as a stationary stochastic process $Y = \{Y(i) : i = 1, 2, 3 \ldots\}$ with constant average μ, where the autocorrelation function has the form

$$C_Y(t) \sim t^{-\alpha}, \quad \text{as} \quad t \to \infty, \tag{10.1}$$

with $0 < \alpha < 1$. The power-law decay of the autocorrelation function implicitly defines a long-range dependent process where $\sum_t C_Y(t) \to \infty$ in the case of an infinite signal. As before, it is possible to define the aggregated process Y^m, and the process Y is called second-order self-similar with Hurst exponent $H = 1 - \alpha/2$ if the autocorrelation function of the aggregated processes are $C_{Y^m}(t) = C_Y(t)$ for all m (Willinger *et al.*, 1996). It is easy to realize that such a process has a variance for the sample average (the aggregated process) that decreases more slowly than the inverse of the sample size. More precisely, $\sigma_m \sim m^{-\alpha}$, thus explaining the enhanced persistence of noise in self-similar signals. Indeed, the scaling form of the variance is one of the techniques used to evaluate and characterize the self-similar properties of time series. Another very common way to characterize and distinguish self-similar signals refers to the power spectrum density, defined as

$$S_Y(f) = \left| \sum_t C_Y(t) \exp\left(i2\pi ft\right) \right|^2, \tag{10.2}$$

that quantifies the level of correlations on a time scale $\tau \sim 1/f$. Generally, $S_Y(f)$ and $C_Y(t)$ are connected by the Wiener–Kintchine relations (McDonald, 1962). For a stationary self-similar processes it is possible to show that $S_Y(f) \sim f^{-\beta}$ with $\beta = 1 - \alpha$, giving rise to what is known as $1/f$ noise.[3]

The characterization and mathematical construction of self-similar processes traces back to the early work of Mandelbrot (1969) and is obviously related to fractal and scale-free phenomena (Mandelbrot, 1982). Also in this case, the heterogeneity of the traffic is indeed characterized by power-law behavior and the corresponding divergence of the statistical fluctuations. The Internet traffic is therefore scale-free and in particular has several correspondences with many physical phenomena, ranging from surface growth (Barabási and Stanley, 1995) to economics (Mandelbrot, 1997; Mantegna and Stanley, 1999) or geophysics (Rodriguez-Iturbe and Rinaldo, 1997). Moreover, the scale-free behavior is not only observed in traffic series. The traffic self-similarity is also affecting network performance and queueing time (Park, Kim, and Crovella, 1997),

---

[3] The $1/f$ noise relates to an exponent $\beta = 1$, but it has become a common nomenclature for all noisy signals with a power spectrum with $\beta \neq 1$.

inter-arrival time, and end-to-end delay[4] that show statistical distributions with heavy tails and power spectrum of $1/f$ type (Csabai, 1994; Paxson and Floyd, 1995; Fowler, 1999; Claffy, 1999). Finally, self-similarity is a general property of the large majority of traffic generated on the Internet, including WWW traffic from file requests on web servers (Crovella and Bestavros, 1997).

10.1.2 Explaining self-similar traffic

The ubiquitous self-similar features of Internet traffic do not imply that all time series are identical. Different locations or types of traffic yield obviously different parameters μ, σ_m, and α, allowing for different levels of burstiness of the time series. However, the failure of the Poisson modeling has been a constant over a decade of measurements and has stimulated intense activity aimed at understanding the origin of this phenomenon.

A first and successful explanation of the self-similar traffic is based on the construction of self-similar processes by using the method of Mandelbrot (1969) and its successive developments (Cox, 1984). In very simple terms, the construction is based on a signal that is due to data transfer sessions whose arrival is modeled by a Poisson process, while their size is power-law distributed. These two ingredients generate aggregate traffic made up of the contribution of all the sessions that are self-similar. Translated into the network language, the aggregate link traffic is the sum of all the contributions during the period of interest of the various sessions such as FTP, HTTP, Telnet etc., at the application level. Indeed, by inspecting the size distribution of sessions on the Internet, clear power law behaviors are found. This fact supports the explanation in terms of the statistical properties of the individual sessions that form the aggregate link traffic (Park, Kim, and Crovella, 1996; Crovella and Bestavros, 1997; Feldmann, Gilbert, Willinger, and Kurtz, 1998; Willinger *et al.*, 2002). In addition, the TCP adaptive mechanism for congestion control may sustain the long-range correlations of traffic in distant areas of the Internet, sustaining and propagating the self-similarity generated in local bottlenecks by the file size distribution (Park *et al.*, 1996; Veres, Kenesi, Molnár, and Vattay, 2000). This picture would easily account for the ubiquitous presence of self-similar properties since it does not depend on the specific traffic technology or network. Rather, it is due to the heavy tailed distribution of transmitted files or Web documents. In this respect, it is also reasonable that self-similarity is not going to disappear with the advent of new technological developments since it is mostly linked to the way users store and organize data.

---

[4] Delays are usually characterized by the round-trip-times (RTT), see Section 10.2.

A different scenario for the origin of self-similarity in Internet traffic has been recently proposed by invoking the presence of a phase transition phenomenon (Ohira and Sawatari, 1998; Takayasu, Fukuda, and Takayasu, 1999; Fukuda, Takayasu, and Takayasu, 2000; Solé and Valverde, 2001; Valverde and Solé, 2002; Guimerà, Arenas, Díaz-Guilera, and Giralt, 2002). In these attempts, the Internet is modeled as a network whose vertices are assumed to be hosts that generate a certain amount of traffic or routers that forward packets. These packets travel to their destination following routing mechanisms that involve a nearest neighbor passing of packets. Different packet dynamics have been considered, with each vertex being able to handle only a finite amount of packets at each time step. Each vertex is also supposed to a have a buffer in which packets can be stored if they arrive in a quantity larger than the handling limit. In this configuration the network's behavior strongly depends upon the injection rate p of new packets. In general, the system undergoes a phase transition from a free phase, in which the number of packets in the network is steady, to a congested phase in which the number of packets is increasing with time (or, in the case of a finite buffer, they saturate the system). At the critical point p_c the system starts to develop long-range dynamical correlation that give rise to the self-similarity of traffic. This self-similar character is usually observed in the $1/f$ noise spectrum of the time series representing the number of packets stored in the vertex buffers. The possibility of a phase transition is confirmed for different delivery dynamics and for different network topologies.

The phase transition mechanism appears as an elegant explanation, that finds its origin as a complex emergent phenomenon due to the global dynamics of the network. In this sense it is opposite to the mechanism based on the heavy tailed distribution of files, that is rather calling for an external feature due to the way people tend to organize data. However, while this latter mechanism can be validated by the study of file and Web document distributions, the phase transition scenario does not yet find clear-cut support from experimental data.[5]

10.2 The global behavior of the Internet traffic and performance

The majority of studies on Internet traffic focus on the local fluctuations and behavior of a single link, host, or router, and in some cases on a small collection of them. A different point of view, however, concerns the study of traffic and performance on a global scale, i.e. for the Internet as a whole. In this case the object of interest does not reside in the characterization of the fluctuations of a single link traffic. Rather, the aim is the quantification of the differences in traffic and performance over a large number of routers and links. Once more, this kind of study looks for

---

[5] An interesting discussion of the models' validation issue is provided by Willinger *et al.* (2002).

a characterization of the large-scale heterogeneity over the global network. This is a very ambitious task, struggling with the huge complexity of the Internet. As we have seen throughout the book, even the characterization of the bare connectivity structure is a very difficult task that has not yet been completely carried out. A global characterization of Internet traffic and performance adds a new dimension to the problem that requires a large amount of resources and the solution of several technical problems. Traffic and performance analysis on a global scale implies the collection of traffic traces and performance estimators that just on a single link represent a noticeable volume of data. It is easy to realize that gathering data on hundreds or thousand of links and routers poses great problems in data handling and storage, as well as in the traffic generated by the measurement process itself. In addition, on the global level these data must be correlated with the detailed physical and geographical structure (bandwidth, link length, etc.) of the Internet.

Despite the technical difficulties, an increasing body of work focuses on the Internet as a whole, especially aimed at forecasting future performance trends. For instance, interdomain traffic can be studied on a global level by looking at data representing all the traffic received by specific service providers (Uhlig and Bonaventure, 2001). In this case it is possible to aggregate all the packets received on the basis of their source address and obtain *prefix* and AS traffic flows. The prefix flow corresponds to the number of packets originating from the same prefix address, while the AS flow is the aggregation of the packets whose source is in the same AS as obtained by the BGP tables of the destination ISP. These data can be correlated with the BGP tables of the studied ISP, providing information on how traffic is distributed in the ISP addressable space. Interestingly, Uhlig and Bonaventure (2001) find that 5–10% of the originating ASs or prefixes are responsible for 90% of the traffic observed in the studied ISP. This readily implies a very heterogeneous traffic distribution. This heterogeneity is also present in the extensive studies performed by the CAIDA measurement infrastructure (see Chapter 3), that allow the construction of traffic matrices representing the traffic flow between pairs of ASs (Claffy, 1999; Huffaker *et al.*, 2000).[6] In this case traffic flows are aggregated on the basis of the source and destination addresses. These traffic matrices can also be correlated with the geographical location of traffic sources and destinations in order to obtain information on regional or country policies and connectivity. For instance, the country matrix constructed from a United States peering location shows a large amount of traffic with source and destination countries different from the United States. This is an indication that United States acts as a hub in handling communications and at the same time that Internet traffic is highly delocalized (Claffy, 1999).

---

[6] Notice that traffic matrices may not be symmetric.

Along with traffic measurements, increasing attention has been devoted to the evaluation of Internet performance at the global level. Indeed various projects have started the collection of performance data from among a large number of source–destination pairs (see Section 3.6). In general these pairs are composed of hosts belonging to universities or research centers; they are connected to many different networks and backbones and have a very wide geographical distribution, so they are likely to represent a statistically significant sample of the Internet as a whole. For each pair the following metrics are usually considered: the geographic distance between the hosts d, the packet loss rate r (the percentage of ICMP packets that do not reach the target point), and the end-to-end delay measured as the RTT (round-trip-time). These measurements are routinely repeated several times each day for a total duration that in some cases spans several years. In some more recent measurement projects the data are gathered with very high frequency, up to every minute, allowing to define quantities such as dayly, weekly, or monthly minimum and average round-trip times RTT_{min} and RTT_{av}, respectively. These data offer the opportunity to test various hypotheses on the statistical behavior of Internet performance.

It is easy to recognize that the RTT is governed by several factors, some of them not related to network performance, but being determined by physical constraints. First, digital information travels along fiber optic cables at almost exactly two-third the speed of light in a vacuum. This gives the mnemonically very convenient value of 1ms RTT per 100 km of cable. Using this speed one can express the geographic distance d in light-milliseconds, obtaining an absolute physical lower bound on the RTT between sites. The actual measured RTT is (usually) larger than this value because of several factors. First, data packets often follow rather circuitous paths leading them through a number of nodes that are far from the geodesic line between the endpoints. Furthermore, each link in a given path is itself far from being straight, often following highways, railways, or power lines (Bovy, Mertodimedjo, Hooghiemstra, Uijterwaal, and Mieghem, 2002). The combination of these factors produces a purely geometrical enhancement factor of the RTT. In addition, there is a minimum processing delay introduced by each router along the way, of the order of 50–250 μs per hop on average, summing up to a few ms for a typical path (Bovy *et al.*, 2002). This can be significant for very close site pairs, but is negligible for most of the paths separated by large distances. On top of this, the presence of cross traffic along the route can cause data packets to be queued in the routers. When the traffic reaches congestion level, the queueing time becomes a very significant part of the RTT and packet loss also sets in. From this perspective, average values of the RTT over one-month periods provide an indication of the level of congestion on the link. Indeed, it is plausible that even on rather congested links there will be a moment in the course of a month when queuing time is negligible, so RTT_{min}

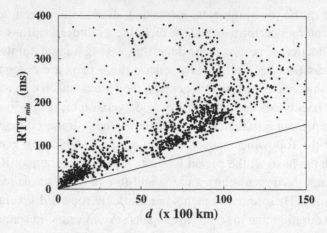

Fig. 10.2 RTT$_{min}$ between 2,114 host pairs (PingER data set of February 2002) as a function of their distance d. Each point corresponds to a different host pair. The line indicates the physical lower bound provided by the speed of light in transmission cables. It is possible to observe the very large fluctuations in the RTT$_{min}$ of different host pairs separated by the same distance. Data from Percacci and Vespignani (2003).

can be taken as an estimate of the best possible communication performance on the given data path, subject only to the intrinsic geometrical enhancement factor and the minimum processing delay. On the contrary, RTT$_{av}$ for a given site pair is obtained by considering the average RTT over one-month periods. This takes into account also the average queueing delay and gives an estimate of the overall communication performance on the given data path.

A preliminary indication of the level of correlation between geographic distance and the end-to-end delay of source–destination pairs is reported in Figure 10.2. The plot represents the obtained relationship for the monthly RTT$_{min}$ of 2,114 host pairs compared with the solid line representing the speed of light in optic fibers at each distance (Huffaker *et al.*, 2000; Lee and Stepanek, 2001; Percacci and Vespignani, 2003). While it is possible to observe a linear correlation of the RTT$_{min}$ with the physical distance of hosts, the data are extremely scattered. Scattered plots of RTT data have qualitatively very similar behavior in all measurements obtained so far (Huffaker *et al.*, 2000; Lee and Stepanek, 2001; Percacci and Vespignani, 2003). Figure 10.2 indicates that end-to-end performance fluctuates conspicuously over the whole range of geographic distances. In particular, the probability distributions of observed RTT exhibit power-law tails (Huffaker *et al.*, 2000). This is again an indication of heterogeneity that, however, still misses a quantitative characterization of the intrinsic fluctuations of performance and their statistical properties. For instance, it is clear that an RTT of 60 ms would be a very

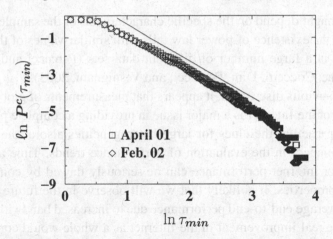

Fig. 10.3 Cumulative distributions of the normalized minimum round-trip-times τ_{min} of the PingER data sets in two different months. The slope of the reference line is -2.0. Data from Percacci and Vespignani (2003).

good performance for a transatlantic link, but a very poor one for a London–Paris connection. In other words, the RTT relative to the actual distance between hosts is a more significant measure of the Internet path than the RTT itself. Indeed, a different characterization of the end-to-end performance is obtained by normalizing the RTT time by the geographical distance between hosts. This defines the absolute performance metrics $\tau = \text{RTT}/d$ that represents the end-to-end delay for unit distance, i.e. the inverse of the overall communication velocity.[7] This metric, and analogously $\tau_{min} = \text{RTT}_{min}/d$ and $\tau_{av} = \text{RTT}_{av}/d$, which represent the minimum and average round trip time for unit distance respectively, allow the comparison of performance between pairs of hosts with different geographical distances.

A method to characterize the level of fluctuations in RTT data is represented by the probability $P(\tau)$ that a pair of hosts present a given value of τ. Significant results are obtained for τ_{min} and τ_{av}, where it is found that data define heavy tailed distributions with power-law behavior (see Figure 10.3). Further evidence of large fluctuations in Internet performance is provided by the analysis of the packet loss data. Also in this case, the probability $P(r)$ that a certain rate r of packet loss occurs on any given pair is well approximated by the power-law behavior $P(r) \sim r^{-\gamma}$ with $\gamma = 1.2 \pm 0.2$. These findings highlight once more the high level of heterogeneity of the Internet. Since we face scale-free distributions, the global average over all the site pairs of $\langle \tau_{min} \rangle$ and $\langle \tau_{av} \rangle$ do not represent characteristic behavior of the whole network. This fact is confirmed by the analysis of different data sets. Although the RTT mean value and packet loss rate across large regions of

---

[7] If d is measured in light-milliseconds τ is a dimensionless variable.

the network might depend on the specific characteristics of the sample (size, world region, etc.), the existence of power law tails with similar values of the exponents is confirmed in a large number of different data sets (Percacci and Vespignani, 2003; Carbone, Coccetti, Dini, Percacci, and Vespignani, 2003).

From the previous discussion, it appears that measurements of global traffic and performance of the Internet is a major issue in providing a complete picture of this network. In particular the clues for large heterogeneities also at the global level have a clear impact in the evaluation of performance trends. Time and scale extrapolation for Internet performance can be seriously flawed by considering just the average properties. It is likely that we will observe in the future an improvement in the average end-to-end performance due to increased bandwidth and router speed, but the real improvement in the Internet as a whole would correspond to a reduction in the huge statistical fluctuations observed nowadays. In this respect, a more complete characterization of global Internet traffic may be considered as a primary goal of measurement projects. On a more theoretical side, the formulation of Internet models able to cast the behavior of Internet performance in the topological structure appears challenging to say the least.

10.3 Internet stability and congestion

A different point of view on global Internet performance is provided by the study of Internet stability to failures. In Chapter 6 the topological resilience of the Internet to link or router outages has been analyzed. In a more accurate picture of the network functioning, however, damages are not static independent events. On the contrary, Internet outages usually stem from failures that propagate through the network. These instabilities have several origins, including router configuration errors, physical outages, or software bugs. In many cases they are due to human errors, and the lack of central coordination in the network often does not allow timely recovery or intervention.

The instability propagation mechanism is referred to as a *route-flap storm*. An overloaded or non-functioning router is marked as unreachable by its BGP peers, which transmit this information to the network and choose alternative paths for routing packets. This changing of route might generate congestions and other BGP routers could fail in maintaining the required interval in the keep-alive transmission (see Section 2.4), being on their turn unreachable. In addition, recovering routers start to download again their peers' BGP tables, loading the network with appreciable traffic. This increased load might cause other congestions, and the total outcome of the process is to initiate a storm or avalanche of outages that can involve extended sections of the Internet. Indeed, several case studies concerning major route-flap storms on the Internet have been reported in the literature (Labovitz *et al.*, 1999; Labovitz and Malan, 1998).

In this respect, Internet failures are very complex dynamical processes where the routing dynamics and policy, the network topology, and the human factor all play together a definite role. These characteristics make the study of Internet instabilities on a large-scale an extremely difficult task that has been approached by a few studies (Paxson, 1997; Labovitz *et al.*, 1998; Labovitz *et al.*, 1999; Magnasco, 2000). In general, the problem is attacked by studying the routing protocol updates. As we have seen in Section 2.4, routers communicate by exchanging update messages about the changes in routes to prefixes or ASs. These messages, that have a time stamp, announce route failures, router unreachability or policy fluctuations, and may be used to estimate the frequency of outages and path withdrawals. Data collection may be carried out at both the inter-domain level (BGP) or at the intra-domain level (IGP).

Since these messages propagate on the network, just a passive measurement from a single peer may provide in principle information on the global occurrence of Internet outages. However, as we discussed for the measurement of the Internet structure, data from a single source cannot provide a full coverage of the whole addressable space. In addition, it is difficult to correlate failure messages in order to decide if they represent a single rout-flap storm or distant, independent events. Despite these difficulties, case studies may give a first picture and some significant indications on Internet instabilities. For instance, Table 10.1 reports the percentage of outages of a certain type occurring in a regional Internet provider in a one year time window as from the study of Labovitz *et al.* (1999). From this table it is clear that the majority of failures registered within the provider are due to software or hardware problems.

More quantitative information can be gathered by studying the time series of routing updates. In particular, it is possible to analyze the power spectrum of these signals and find the main frequency components, highlighting time cycles in the signal. Very interestingly, the IGP and the BGP routing information present two different spectra (Labovitz *et al.*, 1999). While the IGP has a quite flat spectrum with no main components, the BGP spectrum has significant components related to 24 hour and seven day cycles, corresponding to the observation of a low amount of failure to update messages over the week-end and early in the morning.[8] This basic difference implies that the IGP level outages are mainly related to hardware and software problems, and have therefore a strong random component. At the BGP level, however, the main source of instability is associated to high traffic

---

[8] The peers under study were located in the US and therefore the dayly cycle refers to US business hours.

Outage type	Incidence (%)
Maintenance	16.2
Power outage	16.0
Carrier problem/fiber cut	15.3
Unreachability	12.6
Hardware failure	9.0
Interface down	6.2
Routing problems	6.1
Congestion	4.6
Malicious attack	1.5
Software problem	1.3
Miscellaneous	5.9
Unknown	5.6

Table 10.1 *Type and percentage of Internet outages for a regional Internet provider in a one year time window (1997–1998). Maintenance refers to scheduled or emergency upgrades of software and hardware. Malicious attacks indicate failures due denial-of-service or virus attacks (Data from Labovitz* et al., *1999)*

spikes, and therefore the most significant BGP instability stems from congestion collapses. In other words, failures occurring in providers are the initiators of route-flaps that affect more frequently the inter-domain level during usage peak times, when congestion collapses are easily propagated.

Studies have also focused on congestions in single links, the availability of paths, and the mean-time to failures (Paxson, 1997; Labovitz and Malan, 1998; Labovitz *et al.*, 1999). More difficult is instead the measurement of the extension of Internet instabilities and the number of affected users. For these measurements more global analyses are needed. Nevertheless, signatures of global burstiness and scale-free behavior are observed in some time distributions of instability announcements (Magnasco, 2000).

Despite the very partial picture concerning Internet instabilities, there have been several recent attempts to model the occurrence of failure avalanches (Holme and Kim, 2002; Holme, 2002; Motter and Lai, 2002; Moreno, Pastor-Satorras, Vázquez, and Vespignani, 2003). These approaches are based on statistical physics models in which the complex structure of the network connectivity interacts with the traffic load poised on top of edges and vertices. Overloaded elements fail and the network redistributes the traffic load following some predefined dynamical re-routing rules. The failure of an element can trigger additional failures by over-loading other elements because of the traffic load redistribution, in a process that may eventually lead to failure avalanches. The traffic load and its redistribution is intuitively a major ingredient that might spontaneously originate congestion collapse. The lack of precise information on the global traffic load of the Internet has led to the development of models in which the betweenness centrality is considered as an estimate of the actual load. While this is a zero-order assumption, we know that the actual picture is much more complicated. In addition, models have very few inputs on technical heterogeneity (bandwidth, etc.) and the rerouting dynamics, that is not always a shortest path routing policy. Nevertheless, the various modeling assumptions generally lead to the occurrence of avalanche behavior and interesting transitions to congested phases. Unfortunately, at the present stage the comparison with real data is not possible and a validation of the various modeling assumptions is not possible.

Needless to say, the problem of Internet stability and the measurement of its performance is a major issue for the future development of this network. The experimental characterization of these features on a global level is therefore a very relevant path of research that represents the next step in our understanding of the Internet's evolution and structure.

11

Outlook

The Internet is such a rapidly evolving system that sometimes the effort put in its study might appear as extenuating and futile as Sisyphus' pursuit. Indeed, one cannot but ask the question: Would the picture and results obtained today still be valid for the Internet of tomorrow?

The scientific community has just started to partially understand the complex properties of present Internet maps, as new and more ambitious mapping projects are being developed, and new kinds of networks are making their appearance, posing new theoretical and experimental challenges. For instance, P2P and *ad-hoc* networks define a new class of dynamical system for which new types of monitoring infrastructures and different modeling frameworks need to be developed.

Another rapidly changing aspect of the Internet is related to traffic load and performance. The traffic in the Internet is steadily increasing, due to users' demands, along with the available bandwidth of communications lines. Unfortunately, we have not yet achieved a proper theoretical understanding of the interplay and mutual feedback between topological features and traffic load. This makes it very difficult to forecast if the existing infrastructures, as well as new networks joining the Internet, will reorganize in view of this interplay, providing a different global picture of the Internet.

As if all that were not enough, new technical improvements, at the software, hardware, and protocol levels, are constantly changing many of the mechanisms that rule the fundamental processes that make the Internet work at the "microscopic" level. The capacity of routers and computers is constantly increasing, and routing protocols and transport technology are ever evolving. Even the basic IP addressing system is undergoing a major transformation to accommodate more addresses by implementing the new IPv6 protocol.

In such a setting it is natural to wonder if the Internet study may cope with the fast pace of its evolution. However, it is exactly from this perspective that we believe large-scale studies of the Internet are extremely valuable. While the elements and technologies of which the Internet is made are constantly changing, we can reasonably think that the large scale-properties of the resulting network will be preserved in the future. This belief is not unmotivated but actually derives from the lesson we learned from statistical physics.

The study of matter has taught us that although many materials or physical systems are very different at the microscopic level, their large-scale behavior is essentially identical. For instance, a liquid-gas transition might appear as very different from a change in the magnetic properties of ferromagnetic materials; however, both transitions are described by the same statistical laws and show the same phase diagram. More surprisingly, when emergent phenomena take over at the critical point, both systems exhibit the same scaling behavior, even at a quantitative level. This feature has been named "universality" and its theoretical understanding is one of the greatest achievements of modern statistical physics.

Universality stems from the fact that when emergent cooperative phenomena set in, the large-scale behavior is essentially determined by the basic symmetry of the interactions and dynamics of the system's components, while other microscopic details become irrelevant. In general, this is a common feature of complex systems with emergent properties. Naturally, the strict notion of universality must be relaxed when we leave the domain of phase transitions. However, qualitative preservation with respect to changes of the very local details of large-scale emerging properties, such as heavy tailed distribution or the absence of characteristic lengths, is a general property of cooperative phenomena.

A large-scale view of the Internet should not make an exception to this general scenario. Emerging properties, as we tried to convey to the reader of this book, are ubiquitous also in the Internet and appear related to the basic organizing principles of the network. This is also confirmed by the very similar topology of networks based on different technologies such as the physical Internet, the WWW, P2P systems, and other virtual networks. The statistical physics perspective focuses exactly on these basic principles, looking at the possible degree of similarity and, possibly, of universality present in the large-scale properties of the Internet. In this sense, the statistical physics approach acquires a particular value being focused on the properties that are likely to be preserved, despite the continuous technological changes.

A large-scale view, however, does not imply that the study of local and technological properties is a minor or less valuable perspective. On the contrary, a detailed understanding of the Internet's workings at the microscopic level is what makes the Internet evolve and develop for the better. A complete view of this network can therefore be obtained only by combining the various approaches in a multilevel study of the Internet. The vista offered by the large-scale study of the Internet may be used to define more accurate models that can be used to validate and improve new technologies; those, in their turn, may be used to obtain new information on the Internet's fabric and the principles that shape it. In this respect, the experimental effort assumes a fundamental role. Any theoretical understanding and modeling of the Internet walks hand-in-hand with data collection work. While the path opened up by the new generation of models inspired by statistical physics has likely marked a turning point in the Internet's representation, structure, and traffic, data are especially needed to go beyond a view limited by the Internet's topological layout. Our understanding of the Internet has surely made important steps forward, but more efforts are still needed in order to obtain an accurate characterization of this amazing network.

Appendix 1

Graph theory applied to topology analysis

The natural framework for a correct mathematical description of complex networks is graph theory. The origins of graph theory can be traced back to the pioneering work of Euler to solve the Königsberg bridges problem (Euler, 1736), and has now reached a maturity in which a wealth of results of immediate applicability are useful for the understanding of real complex networks. In this appendix we shall provide a cursory introduction to the main definitions and concepts of graph theory, useful for the analysis of real networks. The main sources followed are the books by Chartrand and Lesniak (1986), Bollobás (1998), and Bollobás (1985), as well as the review articles by Albert and Barabási (2002), Dorogovtsev and Mendes (2002), and Newman (2003), covering more recent aspects.

A1.1 Graphs and subgraphs

An undirected graph G is defined by a pair of sets $G = (\mathcal{V}, \mathcal{E})$, where $\mathcal{V}$ is a non-empty countable set of elements, called *vertices* or *nodes*, and $\mathcal{E}$ is a set of *unordered* pairs of different vertices, called *edges* or *links*. Throughtout the book a vertex is reffered to by its order i in the set $\mathcal{V}$. The edge (i, j) joins the vertices i and j, which are said to be *adjacent* or *connected*.[1] The total number of vertices in the graph (the cardinality of the set $\mathcal{V}$) is denoted as N, the size of the graph. The total number of edges is denoted by E.[2] For a graph of size N, the maximum number of edges is $\binom{N}{2}$. A graph with $E = \binom{N}{2}$, i.e. in which all possible pairs of vertices are joined by edges, is called a *complete N-graph*. Undirected graphs are depicted graphically as a set of dots, representing the vertices, joined by lines between pairs of vertices, that represent the corresponding edges, Figure A1.1(a).

---

[1] In the physics literature it is also common to call connected vertices *neighbors* or *nearest neighbors*.
[2] The mathematical nomenclature is *order* for N and *size* for E (Bollobás, 1998). We will follow, however, the terminology exposed in the text, which is more appealing for the physics community.

(a)   (b)

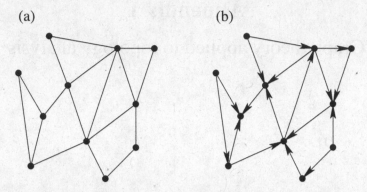

Fig. A1.1   (a) Graphical representation of an undirected graph. The dots represent the vertices. Pairs of adjacent vertices are connected by a line. (b) Graphical representation of a directed graph. Adjacent vertices are connected by arrows, indicating the sense of the corresponding edge.

A directed graph D, or digraph, is defined by a non-empty countable set of vertices $\mathcal{V}$ and a set of *ordered* pairs of different vertices $\mathcal{E}$, that are called directed edges. In a graphical representation, Figure A1.1(b), the ordered nature of the edges is depicted by means of an arrow, indicating the sense of each edge. The main difference between directed and undirected graphs is clearly represented in Figure A1.1. In an undirected graph the presence of an edge between vertices i and j connects the vertices in both directions. However, the presence of an edge from i and j in a directed graph does not necessarily imply the presence of the reverse edge between j and i. This fact has important consequences for the connectedness of a directed graph, as will be discussed in more detail in Section A1.4.

Sometimes, we are also interested in subsets of a graph. A graph $G' = (\mathcal{V}', \mathcal{E}')$ is said to be a *subgraph* of the graph $G = (\mathcal{V}, \mathcal{E})$ if all the vertices in $\mathcal{V}'$ belong to $\mathcal{V}$ and all the edges in $\mathcal{E}'$ belong to $\mathcal{E}$, i.e. $\mathcal{E}' \subset \mathcal{E}$ and $\mathcal{V}' \subset \mathcal{V}$. A *clique* is a complete n-subgraph of size $n < N$.

From a mathematical point of view, it is convenient to define a graph with vertices $\mathcal{V} = \{1, 2, \ldots, N\}$ and edges $\mathcal{E} = \{(i, j)\}$ by means of the *adjacency matrix* $\mathbf{A} = \{A_{ij}\}$. This is a $N \times N$ matrix defined such that

$$A_{ij} = \begin{cases} 1 & \text{if } (i, j) \in \mathcal{E} \\ 0 & \text{if } (i, j) \notin \mathcal{E} \end{cases}. \qquad (A1.1)$$

For undirected graphs the adjacency matrix is symmetric, $A_{ij} = A_{ji}$, and therefore it conveys a great deal of redundant information. For directed graphs, the adjacency matrix is not symmetric.

It is important to note that the above definition of both graphs and digraphs does not allow the existence of *loops* (edges connecting a vertex to itself) nor *multiple edges* (two vertices connected by more than one edge). Graphs with either of these two elements are called *multigraphs* (Bollobás, 1998), and are completely natural in graph theory. We will restrict ourselves, however, to the definitions provided here (no loops nor multiple edges allowed), since these two constructions are most often absent in the systems we are dealing with in this book. The reader should also be warned that throughout this book, and unless stated otherwise, the term "graph" will be used to refer to an undirected graph.

A1.2 Degree and degree distribution

The most basic topological characterization of a graph is given by the degree of its vertices and the relative distribution of degrees, the so-called degree distribution. The degree k_i of a vertex i is defined as the number of edges in the graph incident on the vertex i. While this definition is most obvious for undirected graphs, it needs some refinement for the case of directed graphs. Thus, we define *in-degree* $k_{in,i}$ of the vertex i as the number of edges arriving at i, while its *out-degree* $k_{out,i}$ is defined as the number of edges departing from i. The degree of a vertex in a directed graph is defined by the sum of the in-degree and the out-degree, $k_i = k_{in,i} + k_{out,i}$. In terms of the adjacency matrix, we can write

$$k_{in,i} = \sum_j A_{ji}, \qquad k_{out,i} = \sum_j A_{ij}. \qquad (A1.2)$$

For an undirected graph, with symmetric adjacency matrix, $k_{in,i} = k_{out,i} \equiv k_i$.

When dealing with large graphs from a statistical point of view, or when considering random graphs (see Section 5.1), it is most convenient to characterize them by means of their *degree distribution*. The degree distribution $P(k)$ of an undirected graph is defined as the probability that any randomly chosen vertex has degree k. In the case of directed graphs, one has to consider instead two distributions, the in-degree $P(k_{in})$ and out-degree $P(k_{out})$ distributions, defined as the probability that a randomly chosen vertex has in-degree k_{in} and out-degree k_{out}, respectively. The average degree of an undirected graph is defined as the average value of k over all the vertices in the network

$$\langle k \rangle = \sum_k k P(k) \equiv \frac{2E}{N}, \qquad (A1.3)$$

since each edge end contributes to the degree of a vertex. For a directed graph, the average in-degree and out-degree must be equal

$$\langle k_{in} \rangle = \sum_{k_{in}} k_{in} P(k_{in}) = \langle k_{out} \rangle = \sum_{k_{out}} k_{out} P(k_{out}) \equiv \frac{\langle k \rangle}{2}, \qquad (A1.4)$$

since an edge departing from any vertex must arrive at another vertex. Analogously to the average degree, it is possible to define the nth moment of the degree distribution

$$\langle k^n \rangle = \sum_k k^n P(k). \tag{A1.5}$$

A *sparse* graph has an average degree that is much smaller than the size of the graph, $\langle k \rangle \ll N$.

A1.3 Clustering coefficient

The concept of *clustering*[3] of a graph refers to the tendency observed in many natural networks to form cliques in the neighborhood of any given vertex. In this sense, clustering implies the property that, if the vertex i is connected to the vertex j, and at the same time j is connected to l, then with a high probability i is also connected to l. The clustering of an undirected graph can be quantitatively measured by means of the *clustering coefficient*, introduced by Watts and Strogatz (1998) in the analysis of complex networks. Let us consider the vertex i, with degree k_i, and let us denote by e_i the number of edges existing between the k_i neighbors of i. The clustering coefficient, c_i, of i is defined as the ratio between the actual number of edges among its neighbors, e_i, and its maximum possible value, $k_i(k_i - 1)/2$, i.e.

$$c_i = \frac{2e_i}{k_i(k_i - 1)}. \tag{A1.6}$$

Thus, the clustering coefficient c_i measures the average probability that two neighbors of the vertex i are also connected between them.[4] Given the definition of e_i, it is easy to check that the number of edges among the neighbors of i can be computed in terms of the adjacency matrix $\mathbf{A}$ as (Vázquez *et al.*, 2002a)

$$e_i = \frac{1}{2} \sum_{jl} A_{ij} A_{jl} A_{li}. \tag{A1.7}$$

Therefore, c_i measures the existence of *correlations* in the adjacency matrix, weighted by the corresponding vertex degree. The clustering coefficient of a graph $\langle c \rangle$ is defined as the average value of c_i over all the vertices in the graph, i.e. $\langle c \rangle = \sum_i c_i / N$. Defining the matrix $\mathbf{I}_k$ of components

$$(\mathbf{I}_k)_{ij} = \begin{cases} \delta_{i,j} \dfrac{1}{k_i(k_i - 1)} & \text{if } k_i > 1 \\ 0 & \text{otherwise} \end{cases}, \tag{A1.8}$$

---

[3] Also called *transitivity* in the context of sociology (Wasserman and Faust, 1994).
[4] Note that this measure of clustering has only meaning for $k_i > 1$. For $k_i \leq 1$ we define $c_i \equiv 0$.

where $\delta_{i,j}$ is the Kronecker symbol[5] and k_i the degree of the vertex i, we can write the clustering coefficient in the compact matricial notation

$$\langle c \rangle = \frac{1}{N} \mathrm{Tr}\,(\mathbf{A}^3 \mathbf{I}_k), \qquad (A1.9)$$

where Tr stands for the trace.

In the physical literature sometimes a slightly different definition of the average clustering coefficient is used that leads to simpler calculations without altering the physical meaning (Barrat and Weigt, 2000). According to this definition, the clustering $\langle c' \rangle$ is defined as the fraction of the mean number of edges between the neighbors of a vertex and the mean number of possible edges between those neighbors. In practice, instead of doing the average of the ratio defining originally the clustering coefficient, the coefficient is approximated by the ratio of the averages. This definition corresponds to the concept of the *fraction of transitive triples* used in sociology (Wasserman and Faust, 1994).

A1.4 Connected components and giant component

A very important issue concerning complex networks is the *reachability* of its different vertices, i.e. the possibility of going from one vertex to another following the connections given by the edges in the network. In a connected network every vertex is reachable from any other vertex. The connected components of a graph thus define many properties of its physical structure.

A1.4.1 Component structure in undirected graphs

Let us define a *path* $\mathcal{P}_{i_0,i_n}$ in a graph $G = (\mathcal{V}, \mathcal{E})$ as an ordered collection of $n+1$ vertices $\mathcal{V}_\mathcal{P} = \{i_0, i_1, \dots, i_n\}$ and n edges $\mathcal{E}_\mathcal{P} = \{(i_0, i_1), (i_1, i_2), \dots, (i_{n-1}, i_n)\}$, such that $i_\alpha \in \mathcal{V}$ and $(i_{\alpha-1}, i_\alpha) \in \mathcal{E}$, for all α. The path $\mathcal{P}_{i_0,i_n}$ is said to connect the vertices i_0 and i_n. The *length* of the path $\mathcal{P}_{i_0,i_n}$ is n. A *cycle* is a closed path ($i_0 = i_n$) in which all vertices and all edges are different. A graph is called *connected* if there exists a path connecting any two vertices in the graph. A *component* $\mathcal{C}$ of a graph is defined as a connected subgraph. Two components $\mathcal{C}_1 = (\mathcal{V}_1, \mathcal{E}_1)$ and $\mathcal{C}_2 = (\mathcal{V}_2, \mathcal{E}_2)$ are disconnected if it is impossible to construct a path $\mathcal{P}_{i,j}$ with $i \in \mathcal{V}_1$ and $j \in \mathcal{V}_2$. A *tree* is defined as a connected graph in which the deletion of any edge breaks it into two disconnected components, see Figure A1.2. It is easy to see that, for any tree, $N = E + 1$. A most interesting property of random graphs

---

[5] The Kronecker symbol has a value $\delta_{i,j} = 1$ if $i = j$, and $\delta_{i,j} = 0$ if $i \neq j$.

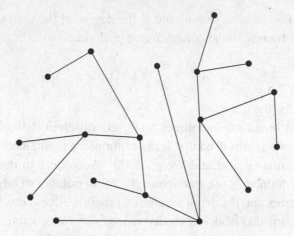

Fig. A1.2   Example of a tree graph.

(Section 5.1) is the distribution of components, and in particular the existence of a *giant component* $\mathcal{G}$, defined as a component whose size scales with the number of vertices of the graph, and therefore diverges in the limit $N \to \infty$. The presence of a giant component implies that a macroscopic fraction of the graph is connected, in the sense that it is possible to find a way across a certain number of edges, joining any two vertices.

A1.4.2 Component structure in directed graphs

The structure of the components of directed graphs is somewhat more complex. The difficulty stems from the obvious fact that, due to the directed nature of the edges, the presence of a path from the vertex i to the vertex j does not necessarily guarantee the presence of a corresponding path from j to i. Therefore, the definition of a giant component becomes more fuzzy.

Following Figure A1.3, the component structure of a directed network can be decomposed into a giant weakly connected component (GWCC), corresponding to the giant component of the same graph in which the edges are considered as undirected, plus a set of smaller disconnected components (DC). In turn, the GWCC is composed by several parts, attending to the directed nature of the edges:

(1) The giant strongly connected component (GSCC), in which there is a directed path joining any pair of vertices.
(2) The giant IN-component (GIN), formed by the vertices from which it is possible to reach the GSCC by means of a directed path.

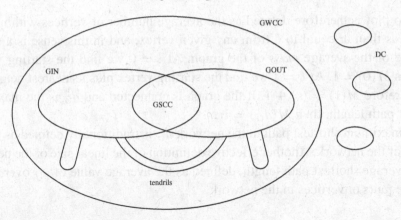

Fig. A1.3 Component structure of a directed graph. Figure adapted from Dorogovtsev *et al.* (2001).

(3) The giant OUT-component (GOUT), formed by the vertices that can be reached from the GSCC by means of a directed path.
(4) The tendrils that contains vertices that cannot reach or be reached by the GSCC (among them, the tubes that connect the GIN and GOUT) that form the rest of the GWCC.

The component structure of directed graphs has important consequences in the accessibility of information in networks such as the World Wide Web (Broder *et al.*, 2000).

A1.5 Shortest path length and betweenness

Even though graphs usually lack a metric, it is possible to define a distance between two vertices i and j, as the number of vertices traversed by the shortest path connecting i and j. This distance, equivalent to the chemical distance usually considered in percolation theory (Bunde and Havlin, 1991), is called the *shortest path length* and denoted as ℓ_{ij}. When two vertices belong to two disconnected components of the graph, we define $\ell_{ij} = \infty$. While it is a symmetric quantity for undirected graphs, the shortest path length ℓ_{ij} does not coincide in general with ℓ_{ji} for directed graphs. There are two main statistical characterizations of the shortest path length. First of all, we may consider the probability distribution $P_\ell(\ell)$ of finding two vertices separated by a distance ℓ. This distribution is related to the so-called *hop plot* $M(\ell)$, defined in the early investigations of the statistical properties of the Internet (Faloutsos *et al.*, 1999):

$$M(\ell) = N \sum_{\ell'=0}^{\ell} P_\ell(\ell'). \qquad (A1.10)$$

The hop plot is therefore defined as the average number of vertices within a distance less than or equal to ℓ from any given vertex, and in this sense is a sort of measure of the average *mass* of the graph. At $\ell = 0$ we find the starting vertex, and thus $M(0) = 1$. At $\ell = 1$ we find the starting vertex plus its nearest neighbors, and therefore $M(1) = \langle k \rangle + 1$. If the graph is connected and d_G is the maximum shortest path length, then $M(d_G) = N$.

The maximum shortest path in the network d_G is traditionally defined as the diameter of the network. Another effective definition of the linear size of the network is the average shortest path length, defined as the average value of ℓ_{ij} over all the possible pairs of vertices in the network

$$\langle \ell \rangle = \sum_\ell \ell P_\ell(\ell) \equiv \frac{2}{N(N-1)} \sum_{i<j} \ell_{ij}. \tag{A1.11}$$

It is worth stressing that the average shortest path length has been also referred in the physics literature as another definition for the diameter of the graph. By definition $\ell_{ij} \leq d_G$, and in the case that the shortest path length distribution is a well behaved and bounded function it is possible to show heuristically that in many cases the two definitions behave in the same way with the network size. In particular, for a regular hypercubic lattice in D dimensions composed by N vertices (or sites), the average shortest path length scales as $\langle \ell \rangle \sim N^{1/D}$; the hop plot, on the other hand, scales as the mass, yielding $M(\ell) \sim \ell^D$. For random graphs, Chapter 5, the average shortest path length grows logarithmically with the size N, $\langle \ell \rangle \sim \log N$, a much slower growth than that found in regular hypercubic lattices. This fact constitutes the so-called *small-world effect*.

Another quantity closely related to the shortest path length is the *betweenness*, sometimes also referred to as *load* or, in sociology, *betweenness centrality* (Newman, 2001b; Goh *et al.* 2001; Brandes, 2001). To go from one vertex to another in the graph, following the shortest path, a certain sequence of vertices is visited. If we count all the vertices visited by the shortest paths between all the possible pairs of vertices in the graph, some key vertices will be visited more often than others. This fact can be quantitatively measured by the betweenness b_i of the vertex i, defined as the total number of shortest paths between any two vertices in the graph that pass through the vertex i. More precisely, if $L_{h,j}$ is the total number of shortest paths from h to j and $L_{h,i,j}$ is the number of these shortest paths that pass through the vertex i, the betweenness is defined as $b_i = \sum L_{h,i,j}/L_{h,j}$, where the sum is over all h, j pairs with $j \neq h$. This definition gives a proper relative weight to shortest paths between pairs of vertices with more than one equivalent shortest path.

In order to characterize statistically this quantity, we can consider the probability distribution $P_b(b)$ that a vertex has betweenness b, and the average betweenness $\langle b \rangle$ is defined as

$$\langle b \rangle = \sum_b b \, P_b(b) \equiv \frac{1}{N} \sum_i b_i. \qquad (A1.12)$$

The betweenness is a magnitude that takes usually values of the order $\mathcal{O}(N)$ or larger. For instance, in a star graph, formed by $N - 1$ vertices with a single edge connected to a central vertex, the betweenness takes a maximum value $N(N - 1)/2$ at the central vertex, and a minimum value $N - 1$ at the peripheral vertices.

Appendix 2

Interface resolution and router topology

Routers have multiple *interfaces*, each one corresponding to a physical connection with another peering router. By definition, each interface is associated with a different IP address. As discussed in Chapter 3, path probing discovers router interfaces. In particular each probing path will record a single interface for each traversed router, generally the one from which the packet arrived to the router. It is therefore possible that probes coming from different monitors, even if directed to the same destination, might enter from different interfaces on the same intermediate router. Thus, each time a router is probed from a different perspective, its interfaces are registered as separate routers. A simple example of the difference between the physical router connectivity and the one registered by probing paths is depicted in Figure A1.1. In this example a network of four physical routers are connected through six different interfaces. The graphs resulting from path probes forwarded from opposite directions are different. By merging these different views, a graph with a false connectivity is obtained and the interfaces (router aliases) resolution becomes indispensable to reconstruct the physical router topology.

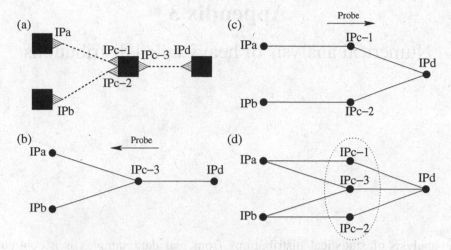

Fig. A2.1 Difference between the physical router topology and the mapped topology caused by the recording of different interfaces on the same router. (a) Physical router topology. To each router (black square), the physical interfaces (filled triangles) correspond to different IP addresses. (b) The graph resulting from a path probe forwarded from right to left. The IP addresses registered and classified as separate vertices are the entering interfaces for the probing packets. (c) Path probes forwarded from left to right yield a different topological view. (d) Merging the two different views creates a false map of the physical connections and inflation in the number of routers.

Appendix 3

Numerical analysis of heavy tailed distributions

The analysis of statistical distributions from real data samples is a continuos struggle with the statistical noise always present because of the inevitably finite sampling or finite size of the system under study. In addition, while statistical distributions often imply averaging over many realizations of the corresponding stochastic process, in most of the cases we can measure just a single realization corresponding to the real world system. These problems are of particular concern in the case of heavy-tailed distributions. In this case, the distribution is spread over a wide range of values with long tails of small but non-negligible probability. The tail, thus, needs of a large number of events to be consistently sampled and data measurements are especially noisy in these regions. As an example, we can consider Figure 4.5, showing the degree frecuency measured in the IR level map obtained with the *Mercator* software (Govindan and Tangmunarunkit, 2000). Despite that the map contains more than 2×10^5 vertices, the probability of the largest degree events is of the order of 10^{-5}. We can therefore expect to detect one or two of these events, as well as none of them. The tail's noise is therefore evident in the last part of the distribution.

Two numerical recipes can be applied in order to smooth the statistical fluctuations generally present in statistical distributions with power-law tails. The first one consists of computing the *cumulative degree distribution* $P^c(k)$, defined by

$$P^c(k) = \sum_{k'=k}^{\infty} P(k').$$ (A3.1)

The function $P^c(k)$ is akin to an integral over k of the original degree distribution, and is therefore smoother than the original function, see Figure A3.1a). For a scale-free network, with $P(k) \sim k^{-\gamma}$ and $\gamma > 1$, we have that $P^c(k) \sim \int_k^{\infty} P(k')\, dk' \sim k^{1-\gamma}$. From a log-log plot of $P^c(k)$ it is possible to compute the slope $1 - \gamma$, and hence the degree exponent, with much higher precision than from the bare degree

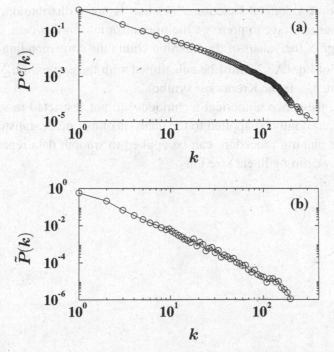

Fig. A3.1   (a) Cumulative degree distribution af the IR Internet map. (b) Binned degree distribution from the same data map with a binning of 30 boxes.

distribution. Moreover, the cumulative distribution allows a clearer identification of a possible cut-off or truncation of the distribution, by providing a smoother representation of the tail.

A second technique refers to the *binning* of data with an exponential bin length for the variable range. Let us consider the explicit example of a scale-free degree distribution. First, the interval of degree values $[m, k_c]$ is discretized in $M + 1$ bins, corresponding to $k_n = m\, r^n$, for $n = 0, 1, \ldots, M$, with $r = (k_c/m)^{1/M}$. Then the binned degree distribution is constructed as $\tilde{P}(k_n)\Delta k_n = \sum_{k=mr^n}^{mr^{n+1}} P(k)\Delta k$. Since the integration intervals are $\Delta k = 1$ and $\Delta k_n = \sum_{k=mr^n}^{mr^{n+1}} 1$ we finally obtain

$$\tilde{P}(k_n) = \frac{\sum_{k=mr^n}^{mr^{n+1}} P(k)}{\sum_{k=mr^n}^{mr^{n+1}} 1}, \qquad \text{for } n = 0, 1, \ldots, M - 1. \tag{A3.2}$$

The function $\tilde{P}(k_n)$ is an integration of the degree distribution, performed over a set of bins of equal size in log-log scale, that preserves the normalization by dividing for the length of the integration intervals. By performing this averaging (or binning), it is possible to smooth out the statistical irregularities observed in the bare numerical degree distribution, see Figure A3.1(b).

In the case of a general function $P(k)$ that is not a distribution, the binning procedure does not have to preserve the normalization. In this case, $\tilde{P}(k_n)$ is just a local average of the values of the function within the corresponding bins and the denominator of Eq. (A3.2) must be substituted with the expression $\sum_{k=mr^n}^{mr^{n+1}} [1 - \delta_{p(k),0}]$, where $\delta_{x,y}$ is the Kronecker symbol.

Obviously, these two numerical techniques are not restricted to scale-free behavior, and can equally be applied to the analysis of any other statistical curve. In particular the binning procedure can be applied to smooth data representation in linear scale by defining linear size bins.

Appendix 4

Degree correlations

A network is said to be *uncorrelated* when the probability that an edge departing from a vertex of degree k arriving at a vertex of degree k' is independent of the degree of the initial vertex k. In undirected networks correlations can be measured by means of the conditional probability $P(k' \mid k)$, defined as the probability that a vertex of degree k is connected to a vertex of degree k'. This function is normalized

$$\sum_{k'} P(k' \mid k) = 1, \tag{A4.1}$$

and is constrained by the degree detailed balance condition (Boguñá and Pastor-Satorras, 2002)

$$k P(k' \mid k) P(k) = k' P(k \mid k') P(k'). \tag{A4.2}$$

Eq. (A4.2) is basically a statement of the conservation of edges among vertices: the total number of edges pointing from vertices with degree k to vertices with degree k' must be equal to the number of edges that point from vertices with degree k' to vertices with degree k. There is a simple way to derive the degree detailed balance condition (Boguñá *et al.*, 2003b). Let us denote by N_k the number of vertices with degree k. Since $\sum_k N_k = N$, we can define the degree distribution as

$$P(k) = \frac{N_k}{N}. \tag{A4.3}$$

To completely define the network, apart from the relative number of vertices of a given degree, we need to specify how the vertices are connected. For this purpose we define the symmetric matrix $E_{kk'}$, that measures the number of edges between vertices of degree k and vertices of degree k' for $k \neq k'$, and two times the number of self-connections for $k = k'$. It is easy to realize that this matrix fulfills the following identities

$$\sum_{k'} E_{kk'} = kN_k, \tag{A4.4}$$

$$\sum_{k} \sum_{k'} E_{kk'} = \langle k \rangle N. \tag{A4.5}$$

In fact, the first relation states that the number of edges emanating from all the vertices of degree k is simply kN_k, while the second indicates that the sum of all the vertices' degrees is equal to two times the number of edges. The identity (A4.5) allows us to define the *joint probability*

$$P(k, k') = \frac{E_{kk'}}{\langle k \rangle N}, \tag{A4.6}$$

where the symmetric function $(2 - \delta_{k,k'}) P(k, k')$ is the probability that a randomly chosen edge connects two vertices of degrees k and k'. The transition probability $P(k' \mid k)$, defined as the probability that an edge from a k vertex points to a k' vertex, can be easily written as

$$P(k' \mid k) = \frac{E_{kk'}}{kN_k} \equiv \frac{\langle k \rangle P(k, k')}{k P(k)}, \tag{A4.7}$$

from where the detailed balance condition Eq. (A4.2) follows immediately as a consequence of the symmetry of $P(k, k')$.

For uncorrelated networks, in which $P(k' \mid k)$ does not depend on k, application of the normalization condition Eq. (A4.1) into Eq. (A4.2) yields the form

$$P(k' \mid k) = \frac{1}{\langle k \rangle} k' P(k') \tag{A4.8}$$

This expression can be easily understood by noticing that the probability that any given edge points to a vertex of degree k is proportional to the density of these vertices times the number of emanated edges. The direct evaluation of $P(k' \mid k)$ is in most real networks a quite difficult task, since the available data, restricted to finite network sizes, yields results extremely noisy and difficult to interpret. For this reason, it is more useful to analyze instead the average degree of the nearest neighbors of the vertices of degree k, defined by (Pastor-Satorras *et al.*, 2001)

$$\bar{k}_{nn}(k) = \sum_{k'} k' P(k' \mid k). \tag{A4.9}$$

If there are no degree correlations, the conditional probability $P(k' \mid k)$ takes the form given in Eq. (A4.8), from which we obtain $\bar{k}_{nn}^0(k) = \langle k^2 \rangle / \langle k \rangle$, independent of k. Therefore, a function $\bar{k}_{nn}(k)$ with an explicit dependence on k signals the presence of degree correlations in the network. Based in the function $\bar{k}_{nn}(k)$, recently a classification framework of natural complex networks in terms of their correlations has been proposed (Newman, 2002a). When $\bar{k}_{nn}(k)$ is an increasing function of k (highly connected vertices tend to connect to highly connected vertices) the network is said to show *assortative mixing*. However, when $\bar{k}_{nn}(k)$ is a decreasing function of k (highly connected vertices prefer to connect with vertices with low degree) the network shows *disassortative mixing*. This classification scheme has proved to be very useful for elucidating the structural properties of complex networks.

In order to provide some explicit example of degree correlations let us focus on the case of growing network models in which new edges always connect new vertices to old vertices. Using general scaling arguments (Dorogovtsev and Mendes, 2003), it is possible to show that the joint probability for a generalized scale-free network with degree distribution $P(k) \sim k^{-\gamma}$ is given by

$$P(k, k') \propto k'^{-(\gamma-1)} k^{-2}, \qquad 1 \ll k \ll k'. \tag{A4.10}$$

Using the degree detailed balance condition, Eq. (A4.2), we can compute the conditional probability as

$$P(k' \mid k) = \frac{\langle k \rangle P(k, k')}{k P(k)} \propto k'^{-(\gamma-1)} k^{-(3-\gamma)}. \tag{A4.11}$$

From here, we obtain in the case of a scale-free network with degree exponent γ

$$\bar{k}_{nn}(k) \propto k^{-(3-\gamma)}, \tag{A4.12}$$

for sufficiently small values of the degree k.

From this last equation we observe that a generalized growing scale-free network exhibits noticeable correlations, as expressed by the functional dependence of $\bar{k}_{nn}$ in k. For the particular case of a Barabási–Albert network, however, the value $\gamma = 3$ renders correlations negligible.

Appendix 5
Scale-free networks: scaling relations

Most natural *scale-free* networks with a power-law tail $P(k) \sim k^{-\gamma}$ in the degree distribution are the outcome of a growth process that, starting from a small core of vertices, increases its size in time by the subsequent addition of new vertices and edges (Albert and Barabási, 2002; Dorogovtsev and Mendes, 2002; Dorogovtsev and Mendes, 2003).

While the overall average degree of scale-free networks with $\gamma > 2$ is constant, the growing nature of these networks is reflected in the scaling of the degree of single vertices as a function of their age, which increases with time as a power-law. Following Section 5.4, consider a growing network model in which a new vertex is added every time step. Thus, if $k_s(t)$ is the average degree of the sth vertex

$$k_s(t) \simeq \left(\frac{t}{s}\right)^{\beta}, \tag{A5.1}$$

where s is the time when the sth vertex was added to the network, t is the age of the network ($s \leq t$), which is proportional to the size N of the network, and β is another characteristic exponent. The exponents γ and β are not independent, but are generally related by the scaling relation (Dorogovtsev and Mendes, 2002)

$$\gamma = 1 + \frac{1}{\beta}. \tag{A5.2}$$

For the case of the Barabási–Albert model, Section 5.5, these exponents take the values $\gamma = 3$ and $\beta = 1/2$, but the relation Eq. (A5.2) is a general property of all scale-free growing network models.

The fact that the network is growing implies that any time we observe it, it is composed of a finite number of vertices $N \sim t$, which through Eq. (A5.1) implies in its turn that the degree of any vertex is bounded by a degree cut-off, $k_c(N)$, that depends on the network size. The presence of the degree cut-off translates in the degree distribution into an explicit dependence on the network size (or time) that we can write in the scaling form as

$$P(k, N) = k^{-\gamma} f\left[\frac{k}{k_c(N)}\right], \tag{A5.3}$$

where $f(x)$ is constant for $x \ll 1$ and decreases very quickly for $x \gg 1$. It is possible to obtain a upper bound for the functional dependence on N of the degree cut-off for generalized uncorrelated random graphs using an extremal theory argument. In the continuous k approximation, consider a random graph with normalized degree distribution in the infinite size limit $P(k) = (\gamma - 1)m^{\gamma-1}k^{-\gamma}$, with $\gamma > 2$ and $k \in [m, \infty[$, where m is the minimum degree of the graph. Consider now that we generate a graph of size N by sorting N independent random variables according to the distribution $P(k)$, obtaining the sample $\{k_1, \ldots, k_N\}$. Let us define K the maximum value of this particular sample, $K = \max\{k_1, \ldots, k_N\}$. When generating an ensemble of graphs, we will obtain in each case a different value of the maximum degree K. Thus, we can define the cut-off $k_c(N)$ as the average value of K, weighted by the distribution $P(k)$. It is easy to see that the probability of this maximum being less than or equal to K is equal to the probability of all the individual values k_i being in their turn less than or equal to K. This means that the distribution function of the maximum value K is just

$$\Pi(K, N) = [\Psi(K)]^N, \tag{A5.4}$$

where $\Psi(K)$ is the distribution function of the probability $P(k)$, i.e. $\Psi(K) = \int_m^K P(k)\, dk = 1 - (K/m)^{-\gamma+1}$. By differentiating Eq. (A5.4), we obtain the probability distribution of maximum values, namely

$$\pi(K, N) = \frac{d\Pi(K, N)}{dK} = \frac{N(\gamma - 1)}{m}\left(\frac{K}{m}\right)^{-\gamma}\left[1 - \left(\frac{K}{m}\right)^{-\gamma+1}\right]^{N-1}. \tag{A5.5}$$

If the degree cut-off is defined as the average value of the maximum K, then we have that

$$k_c(N) = \int_m^\infty K\pi(K, N)\, dK = \frac{Nm}{\lambda}\frac{\Gamma(1 + \lambda)\Gamma(N)}{\Gamma(N + \lambda)}, \tag{A5.6}$$

where $\Gamma(x)$ is the Gamma function and we have defined the constant $\lambda = (\gamma - 2)/(\gamma - 1)$. Using the asymptotic relation $\lim_{N\to\infty} \Gamma(N + a)/\Gamma(N + b) \simeq N^{a-b}$ (Abramowitz and Stegun, 1972), we obtain the leading behavior for large N

$$k_c(N) \simeq m\frac{\Gamma(1 + \lambda)}{\lambda}N^{1-\lambda} \sim mN^{1/(\gamma-1)}. \tag{A5.7}$$

The previous equation is in fact an upper bound for $k_c(N)$, since we have only considered the possible values that the random variables k_i can take, according to the probability distribution $P(k)$. If those values must represent the degree sequence of an actual graph, some constraints would then apply, specially if we want to avoid the presence of loops or multiple edges.

Finally, it is important to stress that the previous result is valid only in the case that the only origin of the degree cut-off resides in the finite number of vertices forming the network. In other situations networks may exhibit a degree cut-off due to external constraints and finite connectivity resources (Amaral *et al.*, 2000; Mossa *et al.*, 2002). In this case the cut-off k_c is not related to the network size and has to be considered as an external parameter (see for instance Sections 4.4, 6.5.1 and 9.4.3).

Appendix 6

The SIR model of virus propagation

The susceptible-infected-removed (SIR) model (Anderson and May, 1992; Murray, 1993; Diekmann and Heesterbeek, 2000) is a classical epidemiological model in which individuals can only exist in three different states: susceptible (healthy), infected, or removed (immunized or dead). On a general network of degree distribution $P(k)$, in which each vertex represents an individual in its corresponding state, the dynamics of this model is defined as follows. At each time step, each susceptible vertex is infected at rate λ, if it is connected to an infected vertex. At the same time, each infected individual becomes removed with probability μ, that, without lack of generality, we set equal to unity.

In order to take into account the heterogeneity induced by the presence of degree fluctuations (Moreno *et al.*, 2002; May and Lloyd, 2001; Newman, 2002b; Boguñá *et al.*, 2003b), let us consider the time evolution of the magnitudes $\rho_k(t)$, $S_k(t)$, and $R_k(t)$, which are the density of infected, susceptible, and removed vertices of degree k at time t, respectively. These variables are connected by means of the normalization condition

$$\rho_k(t) + S_k(t) + R_k(t) = 1. \tag{A6.1}$$

Global quantities such as the epidemic prevalence can be expressed as an average over the various degree classes; for example, we define the total number of removed individuals at time t by $R(t) = \sum_k P(k)R_k(t)$, and the prevalence as $R_\infty = \lim_{t \to \infty} R(t)$. At the mean-field level, for undirected random uncorrelated sparse networks, these densities satisfy the following set of coupled differential equations

$$\frac{d\rho_k(t)}{dt} = -\rho_k(t) + \lambda k S_k(t)\Theta(t), \tag{A6.2}$$

$$\frac{dS_k(t)}{dt} = -\lambda k S_k(t)\Theta(t), \tag{A6.3}$$

$$\frac{dR_k(t)}{dt} = \rho_k(t). \tag{A6.4}$$

The factor $\Theta(t)$ represents the average density of infected individuals of vertices pointed at by any given edge. In uncorrelated networks this quantity can be computed in a self-consistent way. In general, the probability that an edge points to an infected vertex with degree k' is proportional to $k'P(k')$. However, since the infected vertex pointed at by the edge has previously received the disease through an edge that cannot be used for transmission anymore (its originating vertex is removed), the correct expression is proportional to $k-1$, yielding

$$\Theta(t) = \frac{1}{\langle k \rangle} \sum_k (k-1)P(k)\rho_k(t). \tag{A6.5}$$

Equations (A6.2), (A6.3), (A6.4), and (A6.5), combined with the initial conditions $R_k(0) = 0$, $\rho_k(0) = \rho_k^0$, and $S_k(0) = 1 - \rho_k^0$, completely define the SIR model on any random uncorrelated network with degree distribution $P(k)$. We will consider in particular the case of an homogeneous initial distribution of infected individuals, $\rho_k^0 = \rho^0$. In this case, in the limit $\rho^0 \to 0$, we can substitute $\rho_k(0) \simeq 0$ and $S_k(0) \simeq 1$. Under this approximation, Eqs. (A6.3) and (A6.4) can be directly integrated, yielding

$$S_k(t) = e^{-\lambda k \phi(t)}, \qquad R_k(t) = \int_0^t \rho_k(\tau)\,d\tau, \tag{A6.6}$$

where we have defined the auxiliary function

$$\phi(t) = \int_0^t \Theta(\tau)\,d\tau = \frac{1}{\langle k \rangle} \sum_k (k-1)P(k)R_k(t). \tag{A6.7}$$

In order to get a closed relation for the total density of infected individuals, it is more convenient to focus on the time evolution of the averaged magnitude $\phi(t)$. To this purpose, let us compute its time derivative

$$\frac{d\phi(t)}{dt} = \frac{1}{\langle k \rangle} \sum_k (k-1)P(k)\rho_k(t)$$

$$= \frac{1}{\langle k \rangle} \sum_k (k-1)P(k)[1 - R_k(t) - S_k(t)]$$

$$= 1 - \frac{1}{\langle k \rangle} - \phi(t) - \frac{1}{\langle k \rangle} \sum_k (k-1)P(k)e^{-\lambda k \phi(t)}, \tag{A6.8}$$

where we have introduced the time dependence of $S_k(t)$ obtained in Eq. (A6.6). Once Eq. (A6.8) is solved, we can obtain the total epidemic prevalence R_∞ as a function of $\phi_\infty = \lim_{t \to \infty} \phi(t)$. Since $R_k(\infty) = 1 - S_k(\infty)$, we have

$$R_\infty = \sum_k P(k)(1 - e^{-\lambda k \phi_\infty}). \tag{A6.9}$$

For a general $P(k)$ distribution, Eq. (A6.8) cannot be generally solved in a closed form. However, we can still get useful information on the infinite time limit; i.e. at the end of the epidemics. Since we have that $\rho_k(\infty) = 0$, and consequently $\lim_{t \to \infty} d\phi(t)/dt = 0$, we obtain from Eq. (A6.8) the following self-consistent equation for ϕ_∞

$$\phi_\infty = 1 - \frac{1}{\langle k \rangle} - \frac{1}{\langle k \rangle} \sum_k (k - 1) P(k) e^{-\lambda k \phi_\infty}. \tag{A6.10}$$

The value $\phi_\infty = 0$ is always a solution. In order to have a non-zero ϕ_∞ solution, i.e. a prevalence $R_\infty > 0$, the condition

$$\frac{d}{d\phi_\infty} \left(1 - \frac{1}{\langle k \rangle} - \frac{1}{\langle k \rangle} \sum_k (k - 1) P(k) e^{-\lambda k \phi_\infty} \right) \Bigg|_{\phi_\infty = 0} \geq 1 \tag{A6.11}$$

must be fulfilled. This relation implies

$$\frac{\lambda}{\langle k \rangle} \sum_k k(k - 1) P(k) \geq 1, \tag{A6.12}$$

which defines the epidemic threshold

$$\lambda_c = \frac{\langle k \rangle}{\langle k^2 \rangle - \langle k \rangle}, \tag{A6.13}$$

below which the epidemic prevalence is $R_\infty = 0$, and above which it attains a finite value $R_\infty > 0$. It is interesting to notice that this is precisely the same value found for the percolation threshold in generalized networks (see Section 6.5). This is hardly suprising since, as stressed by Grassberger (1983), the SIR model can be mapped to an edge percolation process. The SIR model is thus no exception to the general absence of an epidemic threshold in networks with diverging degree fluctuations, i.e. $\langle k^2 \rangle \to \infty$. The present results are valid for infinite size networks. In the case of finite networks of size N the usual size corrections set in.

For the case of correlated random networks (Appendix A4) which are completely defined by the degree distribution $P(k)$ and the conditional probability $P(k' \mid k)$ that a vertex of degree k has an edge pointing to a vertex of degree k', it can be proved (Boguñá *et al.*, 2003b) that the epidemic threshold is inversely proportional to the largest eigenvalue $\tilde{\Lambda}_m$ of the matrix

$$\tilde{C}_{kk'} = \frac{k(k'-1)}{k'} P(k' \mid k). \tag{A6.14}$$

In the same way as for the SIS model, the eigenvalue $\tilde{\Lambda}_m$ for scale-free networks with diverging fluctuations and any sort of correlations can be shown to diverge in the thermodynamic limit, provided that the minimum connectivity of the network is larger than one, thus recovering the null epidemic threshold obtained in the uncorrelated case. In the case of a minimum connectivity equal to one a few exceptions might occur as reported by Vázquez and Moreno (2003) and Boguñá *et al.* (2003b).

References

Abbate, J. (2000), *Inventing the Internet*, Cambridge, Massachusetts: MIT Press.

Abramowitz, M. and Stegun, I. A. (1972), *Handbook of Mathematical Functions*, New York: Dover.

Abramson, G. and Kuperman, M. (2001), "Small world effect in an epidemiological model," *Phys. Rev. Lett.* **86**, 2909–2912.

Adamic, L. A. (1999), "The small world Web," in S. Abiteboul and A.-M. Vercoustre (eds.), *Proc. 3rd European Conf. Research and Advanced Technology for Digital Libraries, ECDL*, Springer-Verlag, pp. 443–452.

— (2001), "Network dynamics: the World Wide Web," Ph. D. Thesis, Applied Physics Department, Stanford University.

Adamic, L. and Adar, E. (2001), "Friends and neighbors on the Web," citeseer.nj.nec. com/380967.html

Adamic, L. A. and Huberman, B. A. (2001), "The Web's hidden order," in *Communications of the ACM*, Vol. 44, New York, NY: ACM Press, pp. 55–60.

Adamic, L. A., Lukose, R. M., and Huberman, B. A. (2003), "Local search in unstructured networks," in S. Bornholdt and H. G. Schuster (eds.), *Handbook of Graphs and Networks: From the Genome to the Internet*, Berlin: Wiley-VCH, pp. 295–317.

Adamic, L. A., Lukose, R. M., Puniyani, A. R., and Huberman, B. A. (2001), "Search in power-law networks," *Phys. Rev. E* **64**, 046135.

Adar, E. and Huberman, B. (2000), "Free riding on Gnutella," *First Monday* **5**, 10.

Aggarwal, C., Al-Garawi, F., and Yu, P. (2001). "Intelligent crawling on the World Wide Web with arbitrary predicates." in Proceedings of 10th International World Wide Web Conference, pp. 96–105.

Aiello, W., Chung, F., and Lu, L. (2001), "A random graph model for power law graphs," *Experimental Math.* **10**, 53–66.

Albert, R. and Barabási, A.-L. (2000), "Topology of evolving networks: local events and universality," *Phys. Rev. Lett.* **85**, 5234–5237.

— (2002), "Statistical mechanics of complex networks," *Rev. Mod. Phys.* **74**, 47–97.

Albert, R., Jeong, H., and Barabási, A.-L. (1999), "Diameter of the World-Wide Web," *Nature* **401**, 130–131.

— (2000a), "Error and attack tolerance of complex networks," *Nature* **406**, 378–382.

— (2000b), "Error and attack tolerance of complex networks (correction)," *Nature* **409**, 542.

Alvarez-Hamelin, J. I. and Schabanel, N. (2003), "Topologie d'Internet, modeles, graphes aleatoires, algorithmique," in 4èmes rencontres francophones sur les Aspects Algorithmiques des Télécommunications (AlgoTel 2003), INRIA 75–82.

253

Amaral, L. A. N., Scala, A., Barthélémy, M., and Stanley, H. E. (2000), "Classes of small-world networks," *Proc. Natl. Acad. Sci. USA* **97**, 11149–11152.

Anderson, R. M. and May, R. M. (1992), *Infectious Diseases in Humans*, Oxford: Oxford University Press.

Arenas, A., Díaz-Guilera, A., and Guimerà, R. (2001), "Communication networks with hierarchical branching," *Phys. Rev. Lett.* **86**, 3196–3199.

Aron, J. L., O'Leary, M., Gove, R. A., Azadegan, S., and Schneider, M. C. (2002), "The benefits of a notification process in addressing the worsening computer virus problem: results of a survey and a simulation model," *Computers and Security* **21**, 142–163.

Bailey, N. T. J. (1975), *The Mathematical Theory of Infectious Diseases*, 2nd edn, London: Griffin.

Barabási, A. L. (2002), *Linked: The New Science of Networks*, Cambridge: Perseus Publishing.

Barabási, A.-L. and Albert, R. (1999), "Emergence of scaling in random networks," *Science* **286**, 509–511.

Barabási, A.-L., Albert, R., and Jeong, H. (1999), "Mean-field theory for scale-free random networks," *Physica A* **272**, 173–187.

— (2000), "Scale-free characteristics of random networks: the topology of the World-Wide Web," *Physica A* **281**, 69–77.

Barabási, A.-L. and Stanley, H. E. (1995), *Fractal Concepts in Surface Growth*, Cambridge: Cambridge University Press.

Baran, P. (1964), "On distributed communications networks," *IEEE Transactions of the Professional Technical Group on Communications Systems* **CS-12**, 1–9.

Barford, P., Bestavros, A., Byers, J., and Crovella, M. (2001), "On the marginal utility of deploying measurement infrastructure," in Proceedings of the ACM SIGCOMM Internet Measurement Workshop 2001, California.

Barrat, A. and Weigt, M. (2000), "On the properties of small-world network models," *Eur. Phys. J. B* **13**, 547–560.

Barrie, J. M. and Presti, D. E. (1996), "The World-Wide-Web as an instructional tool," *Science* **274**, 371–372.

Barthélémy, M. and Amaral, L. A. N. (1999a), "Erratum. Small-world networks: evidence for a crossover picture," *Phys. Rev. Lett.* **82**, 5180.

— (1999b), "Small-world networks: evidence for a crossover picture," *Phys. Rev. Lett.* **82**, 3180–3183.

Bellovin, S. M. (1993), "Packets found on an Internet," *Comput. Commun. Rev.* **23**, 26–31.

Bender, E. A. and Canfield, E. R. (1978), "The asymptotic number of labeled graphs with given degree distribution," *Journal of Combinatorial Theory A* **24**, 296–307.

Bianconi, G. and Barabási, A.-L. (2001), "Competition and multiscaling in evolving networks," *Europhys. Lett.* **54**, 436–442.

Bianconi, G. and Capocci, A. (2003), "Number of loops of size h in growing scale-free networks," *Phys. Rev. Lett.* **90**, 078701.

Binney, J. J., Dowrick, N. J., Fisher, A. J., and Newman, M. E. J. (1992), *The Theory of Critical Phenomena*, Oxford: Oxford University Press.

Boguñá, M. and Pastor-Satorras, R. (2002), "Epidemic spreading in correlated complex networks," *Phys. Rev. E* **66**, 047104.

Boguñá, M., Pastor-Satorras, R., and Vespignani, A. (2003a), "Absence of epidemic threshold in scale-free networks with degree correlations," *Phys. Rev. Lett.* **90**, 028701.

— (2003b), "Epidemic spreading in complex networks with degree correlations," *Lecture Notes in Physics* **625**, 127–147.

Bollobás, B. (1981), "Degree sequences of random graphs," *Discrete Math.* **33**, 1–19.
— (1985), *Random Graphs*, London: Academic Press.
— (1998), *Modern Graph Theory*, New York: Springer-Verlag.
Bollobás, B. and Riordan, O. (2003), "Mathematical results on scale-free random graphs," in S. Bornholdt and H. G. Schuster (eds.), *Handbook of Graphs and Networks: From the Genome to the Internet*, Berlin: Wiley–VCH, pp. 1–34.
Bovy, C. J., Mertodimedjo, H. T., Hooghiemstra, G., Uijterwaal, H., and Mieghem, P. V. (2002), "Analysis of end-to-end delay measurements in Internet," in Proceedings of Passive and Active Measurement (PAM2002), Fort Collins, USA, March 25–27, pp. 26–33.
Brandes, U. (2001), "A faster algorithm for betweenness centrality," *J. Math. Soc.* **25**, 163–177.
Brin, S. and Page, L. (1998), "The anatomy of a large-scale hypertextual Web search engine," *Computer Networks and ISDN Systems* **30**, 107–117.
Broder, A. Z., Kumar, S. R., Maghoul, F., Raghavan, P., Rajagopalan, S., Stata, R., Tomkins, A., and Wiener, J. L. (2000), "Graph structure in the Web," *Computer Networks* **33**, 309–320.
Broido, A. and Claffy, K. (2001), "Internet topology: connectivity of IP graphs," in SPIE International symposium on Convergence of IT and Communication, Denver, CO.
— (2002), "Topological resiliance in IP and AS graphs," http://www.caida.org/analysis/ topology/resilience/index.xml
Bu, T. and Towsley, D. (2002), "On distinguishing between Internet power law topology generators," in Proceedings of INFOCOM, 2002.
Buchanan, M. (2002), *Small World: Uncovering Nature's Hidden Networks*, London: Weidenfeld & Nicolson.
Bunde, A. and Havlin, S. (1991), "Percolation," in A. Bunde and S. Havlin (eds.), *Fractals and Disordered Systems*, Heidelberg: Springer Verlag, pp. 51–95.
Burch, H. and Cheswick, B. (1999), "Mapping the Internet," *IEEE Computer* **32**, 97–98.
Burda, Z., Correia, J. D., and Krzywicki, A. (2001), "Statistical ensemble of scale-free random graphs," *Phys. Rev. E* **64**, 046118.
Bush, B. W., Files, C. R., and Thompson, D. R. (2001), "Empirical characterization of infrastructure networks," Technical Report LA-UR-01-5784, Los Alamos National Laboratory.
Caldarelli, G., Capocci, A., De Los Rios, P., and Muñoz, M. A. (2002), "Scale-free networks from varying vertex intrinsic fitness," *Phys. Rev. Lett.* **89**, 258702.
Caldarelli, G., Marchetti, R., and Pietronero, L. (2000), "The fractal properties of Internet," *Europhys. Lett.* **52**, 386.
Caldarelli, G., Pastor-Satorras, R., and Vespignani, A. (2002), "Cycles and local ordering in complex networks," e-print cond-mat/0212026.
Callaway, D. S., Hopcroft, J. E., Kleinberg, J. M., Newman, M. E. J., and Strogatz, S. H. (2001), "Are randomly grown networks really random?" *Phys. Rev. E* **64**, 041902.
Callaway, D. S., Newman, M. E. J., Strogatz, S. H., and Watts, D. J. (2000), "Network robustness and fragility: percolation on random graphs," *Phys. Rev. Lett.* **85**, 5468–5471.
Calvert, K. L., Doar, M. B., and Zegura, E. W. (1997), "Modeling Internet topology," *IEEE Communications Magazine* **35**(6), 160–163.
Capocci, A., Caldarelli, G., Marchetti, R., and Pietronero, L. (2001), "Growing dynamics of Internet providers," *Phys. Rev. E* **64**, 35105.
Carbone, L., Coccetti, F., Dini, P., Percacci, R., and Vespignani, A. (2003), "The spectrum of Internet performance," in Proceedings of Passive and Active Measurement (PAM2003).

Carlson, J. M. and Doyle, J. (1999), "Highly optimized tolerance: a mechanism for power laws in designed sistems," *Phys. Rev. E* **60**, 1412–1427.

C.E.R. Team (2001), "Denial of service attacks," http://www.cert.org/tech_tips/denial_of_service.html

Chakrabarti, S., Dom, B., Gibson, D., Kleinberg, J., Kumar, R., Raghavan, P., Rajagopalan, S., and Tomkins, A. (1999), "Hypersearching the Web," *Scientific American* **280**, 44–52.

Chakrabarti, S., van den Berg, M., and Dom, B. (1999), "Focused crawling: a new approach to topic-specific Web resource discovery," *Computer Networks* **31**, 1623–1640.

Chang, H., Govindan, R., Jamin, S., Shenker, S., and Willinger, W. (2001), "Towards capturing representative AS-level Internet topologies," Technical Report Technical Report CSE-TR-454-02, EECS Department, University of Michigan.

Chang, H., Jamin, S., and Willinger, W. (2001), "Inferring AS-level internet topology from router-level path traces," in Proceedings of SPIE ITCom 2001, Denver, CO, August 2001.

Cho, J., Garcia-Molina, H., and Page, L. (1998), "Efficient crawling through URL ordering," *Computer Networks* **30**, 1–7.

Chartrand, G. and Lesniak, L. (1986), *Graphs and Digraphs*, Menlo Park: Wadsworth & Brooks/Cole.

Claffy, K. (1999), "Internet measurement and data analysis: topology, workload, performance and routing statistics," http://www.caida.org/outreach/papers/1999/Nae

Claffy, K., Miller, G., and Thompson, K. (1998), "The nature of the beast: recent traffic measurements from an internet backbone," Proceedings INET'98 Geneva, Switzerland.

Claffy, K., Monk, T., and McRobb, D. (1999), "Internet tomography," *Nature*, Web Matters, 7 January.

Cohen, R., ben-Avraham, D., and Havlin, S. (2002a), "Efficient immunization of populations and computers," e-print cond-mat/0209586.

— (2002b), "Percolation critical exponents in scale-free networks," *Phys. Rev. E* **66**, 036113.

Cohen, R., Erez, K., ben Avraham, D., and Havlin, S. (2000), "Resilience of the Internet to random breakdowns," *Phys. Rev. Lett.* **85**, 4626.

— (2001a), "Breakdown of the Internet under intentional attack," *Phys. Rev. Lett.* **86**, 3682–3685.

— (2001b), "Reply to the comment on 'Breakdown of the Internet under intentional attack'," *Phys. Rev. Lett.* **87**, 219802.

Cohen, R. and Havlin, S. (2003), "Scale-free networks are ultrasmall," *Phys. Rev. Lett.* **90**, 058701.

Cox, D. R. (1984), "Long-range dependence," a review in H. A. David and H. David (eds.), *Statistics: An Appraisal*, Iowa State University Press, pp. 55–74.

Crovella, M. E. and Bestavros, A. (1997), "Self-similarity in World Wide Web traffic: evidence and possible causes," *IEEE/ACM Transaction on Networking* **5**, 835–846.

Crovella, M. E., Lindemann, C., and Reiser, M. (1997), "Internet performance modeling," *IEEE/ACM Transaction on Networking* **5**, 835–846.

Crucitti, P., Latora, V., Marchiori, M., and Rapisarda, A. (2003), "Efficiency of scale-free networks: error and attack tolerance," *Physica A* **320**, 622–642.

Csabai, I. (1994), "1/f noise in computer networks traffic," *Journal of Physics A: Math. Gen.* **27**, 417–419.

Dezsö, Z. and Barabási, A.-L. (2002), "Halting viruses in scale-free networks," *Phys. Rev. E* **65**, 055103(R).

Diekmann, O. and Heesterbeek, J. (2000), *Mathematical Epidemiology of Infectious Diseases: Model Building, Analysis and Interpretation*, New York: John Wiley & Sons.

Dill, S., Kumar, S. R., McCurley, K. S., Rajagopalan, S., Sivakumar, D., and Tomkins, A. (2001), "Self-similarity in the Web," *The VLDB Journal*, pp. 69–78.

Doar, M. B. (1996), "A better model for generating test networks," in J. Crowcroft and H. Schulzrinne (eds.), Proceedings of the IEEE Global Telecommunications Conference (Globecom 96), London, pp. 86–93.

Doar, M. and Leslie, I. (1993), "How bad is naïve multicast routing?" in M. G. Hluchyj (ed.), *Proceedings of the 12th Annual Joint Conference of the IEEE Computer and Communications Societies on Networking: Foundations for the Future*, volume 1, Los Alamitos, CA: IEEE Computer Society Press, pp. 82–89.

Dorogovtsev, S. N., Goltsev, A. V., and Mendes, J. F. F. (2002), "Pseudofractal scale-free web," *Phys. Rev. E* **65**, 066122.

Dorogovtsev, S. N. and Mendes, J. F. F. (2000), "Scaling behaviour of developing and decaying networks," *Europhys. Lett.* **52**, 33.

— (2001a), "Comment on 'Breakdown of the Internet under intentional attack'," *Phys. Rev. Lett.* **87**, 219801.

— (2001b), "Effect of the accelerating growth of communications networks on their structure," *Phys. Rev. E* **63**, 025101.

— (2002), "Evolution of networks," *Adv. Phys.* **51**, 1079–1187.

— (2003), *Evolution of Networks: From Biological Nets to the Internet and WWW*, Oxford: Oxford University Press.

Dorogovtsev, S. N., Mendes, J. F. F., and Samukhin, A. N. (2001), "Giant strongly connected component of directed networks," *Phys. Rev. E* **64**, 025101.

— (2000), "Structure of growing networks with preferential linking," *Phys. Rev. Lett.* **85**, 4633–4636.

Ebel, H., Mielsch, L.-I., and Bornholdt, S. (2002), "Scale-free topology of e-mail networks," *Phys. Rev. E* **66**, 035103.

Eckmann, J.-P. and Moses, E. (2002), "Curvature of co-links uncovers hidden thematic layers in the World Wide Web," *Proc. Natl. Acad. Sci. USA* **99**, 5825–5829.

Erdös, P. and Rényi, P. (1959), "On random graphs," *Publicationes Mathematicae* **6**, 290–297.

— (1960), "On the evolution of random graphs," *Publ. Math. Inst. Hung. Acad. Sci.* **5**, 17–60.

— (1961), "On the strength of connectedness of random graphs," *Acta. Math. Sci. Hung* **12**, 261–267.

Essam, J. W. (1980), "Percolation theory," *Rep. Prog. Phys.* **43**, 833–912.

Euler, L. (1736), "Solutio problematis ad geometriam situs pertinentis," *Comment. Academiae Sci. J. Petropolitanae* **8**, 128–140.

Fabrikant, A., Koutsoupias, E., and Papadimitriou, C. H. (2002), "Heuristically optimized trade-offs: A new paradigm for power laws in the Internet," in Proceedings of the 29th International Colloquium on Automata, Languages, and Programming (ICALP), Malaga, Spain.

Faloutsos, M., Faloutsos, P., and Faloutsos, C. (1999), "On power-law relationship of the Internet topology," *Comput. Commun. Rev.* **29**, 251–263.

Feldmann, A., Gilbert, A. C., Willinger, W., and Kurtz, T. G. (1998), "The changing nature of network traffic: scaling phenomena," *Computer Communications Review* **28**, 5–29.

Flake, G., Lawrence, S., and Giles, C. L. (2000), "Efficient identification of Web communities," in Sixth ACM SIGKDD International Conference on Knowledge Discovery and Data Mining, Boston, MA, pp. 150–160.

Fowler, T. B. (1999), "A short tutorial on fractals and Internet traffic," *The Telecommunication Review* **10**, 1–14.

Fukuda, K., Takayasu, H., and Takayasu, M. (2000), "Origin of critical behavior in Ethernet traffic," *Physica A* **287**, 289–301.

Gardiner, C. W. (1985), *Handbook of Stochastic Methods*, 2nd edn, Berlin: Springer.

Gibson, D., Kleinberg, J. M., and Raghavan, P. (1998), "Inferring Web communities from link topology," in UK Conference on Hypertext, pp. 225–234.

Gilbert, E. N. (1959), "Random graphs," *Annals of Mathematical Statistics* **30**, 1141–1144.

Gillies, J. and Cailliau, R. (2000), *How the Web was Born*, Oxford: Oxford University Press.

Gnedenko, B. V. (1962), *The Theory of Probability*, New York: Chelsea.

Goh, K.-I., Kahng, B., and Kim, D. (2001), "Universal behavior of load distribution in scale-free networks," *Phys. Rev. Lett.* **87**, 278701.

— (2002), "Fluctuation-driven dynamics of the Internet topology," *Phys. Rev. Lett* **88**, 108701.

Goh, K., Oh, E., Jeong, H., Kahng, B., and Kim, D. (2002), "Classification of scale-free networks," *Proc. Natl. Acad. Sci. USA* **99**, 12583–12588.

Govindan, R. and Reddy, A. (1997), "An analysis of Internet inter-domain topology and route stability," in Proceedings of IEEE INFOCOM'97, p. 850.

Govindan, R. and Tangmunarunkit, H. (2000), "Heuristics for Internet map discovery," in INFOCOM 2000, *Proceedings of the Nineteenth Annual Joint Conference of the IEEE Computer and Communications Societies*, Volume 3, IEEE Computer Society Press, pp. 1371–1380.

Grassberger, P. (1983), "On the critical behavior of the general epidemic process and dynamical percolation," *Math. Biosci.* **63**, 157–172.

Guillaume, J.-L. and Latapy, M. (2003), "A realistic model for complex networks", e-print cond-mat/0307095.

Guimerà, R., Arenas, A., Díaz-Guilera, A., and Giralt, F. (2002), "Dynamical properties of model communication networks," *Phys. Rev. E* **66**, 026704.

Guimerà, R., Danon, L., Díaz-Guilera, A., Girault, F., and Arenas, A. (2002), "Self-similar community structure in organisations," e-print cond-mat/0211498.

Harley, C. D., Slade, R., Harley, D., Spafford, E. H., and Gattiker, U. E. (2001), *Viruses Revealed*, New York: McGraw-Hill.

Harris, T. E. (1974), "Contact interactions on a lattice," *Ann. Prob.* **2**, 969–988.

Holme, P. (2002), "Edge overload breakdown in evolving networks," *Phys. Rev. E* **66**, 036119.

Holme, P. and Kim, B. J. (2002), "Vertex overload breakdown in evolving networks," *Phys. Rev. E* **65**, 066109.

Huberman, B. A. (2001), *The Laws of the WEB*, Cambridge, MA: MIT Press.

Huberman, B. A. and Adamic, L. A. (1999), "Growth dynamics of the World-Wide Web," *Nature* **401**, 131.

Huberman, B. and Lukose, R. (1997), "Social dilemmas and Internet congestion," *Science* **277**, 535.

Huberman, B., Pirolli, P., Pitkow, J., and Lukose, R. (1998), "Strong regularities in World Wide Web surfing," *Science* **280**, 95.

Huffaker, B., Fomenkov, M., Moore, D., Nemeth, E., and Claffy, K. (2000), "Measurements of the Internet topology in the asia-pacific region," in INET 2000, Yokohama, Japan, 18–21 July, The Internet Society.

Huffaker, B., Fomenkov, M., Moore, D., Plummer, D., and Claffy, K. (2002), "Distance metrics in the Internet," in *IEEE International Telecommunications Symposium (ITS)*, Brazil, Sept 2002, IEEE Computer Society Press.

Huffaker, B., Plummer, D., Moore, D., and Claffy, K. (2002), "Topology discovery by active probing," in Proceedings of 2002 Symposium on Applications and the Internet (SAINT) Workshops, pp. 90–96.

Jamin, S., Jin, C., Jin, Y., Raz, D., Shavitt, Y., and Zhang, L. (2000), "On the placement of Internet instrumentation," in INFOCOM (1), pp. 295–304.

Jeong, H., Mason, S., Barabási, A. L., and Oltvai, Z. N. (2001), "Lethality and centrality in protein networks," *Nature* **411**, 41.

Jeong, H., Néda, Z., and Barabási, A.-L. (2003), "Measuring preferential attachment for evolving networks," *Europhys. Lett.* **61**, 567–572.

Jeong, H., Tombor, B., Albert, R., Oltvai, Z. N., and Barabási, A. L. (2000), "The large scale organization of the metabolic network," *Nature* **407**, 651.

Jin, C., Chen, Q., and Jamin, S. (2000), "INET: Internet topology generators," Technical Report CSE-TR-433-00 EECS Department, University of Michigan.

Jin, S. and Bestavros, A. (2002), "Small-world internet topologies: possible causes and implications on scalability of end-system multicast," Technical Report BUCS-2002-004, Boston University.

Jung, S., Kim, S., and Kahng, B. (2002), "Geometric fractal growth model for scale-free networks," *Phys. Rev. E* **65**, 056101.

Kahng, B., Park, Y., and Jeong, H. (2002), "Robustness of the in-degree exponent for the World Wide Web," *Phys. Rev. E* **66**, 046107.

Kephart, J. O., Sorkin, G. B., Chess, D. M., and White, S. R. (1997), "Fighting computer viruses," *Scientific American* **277**(5), 56–61.

Kephart, J. O. and White, S. R. (1991), "Directed-graph epidemiological models of computer viruses," in Proceedings of the 1991 IEEE computer society symposium on research in security and privacy (SSP 1991), IEEE, Washington, Brussels, Tokyo, pp. 343–361.

— (1993), "Measuring and modeling computer virus prevalence," in Proceedings of the 1993 IEEE computer society symposium on security and privacy (SSP 1993), IEEE, Washington, Brussels, Tokyo, pp. 2–15.

Kephart, J. O., White, S. R., and Chess, D. M. (1993), "Computers and epidemiology," *IEEE Spectrum* **30**, 20–26.

Kim, B. J., Yoon, C. N., Han, S. K., and Jeong, H. (2002), "Path finding strategies in scale-free networks," *Phys. Rev. E* **65**, 027103.

Kleinberg, J. and Lawrence, S. (2001), "The structure of the Web," *Science* **294**, 1849–1850.

Kleinberg, J. M. (1998), "Authoritative sources in a hyperlinked environment," in Proceedings ACM-SIAM Symposium on Discrete Algorithms, San Francisco, pp. 668–677.

— (2000), "Navigation in a small world," *Nature* **406**, 845.

— (2002), "Small-world phenomena and the dynamics of information," in *Advances in Neural Information Processing Systems* 14, Cambridge: MIT Press.

Kleinberg, J. M., Kumar, R., Raghavan, P., Rajagopalan, S., and Tomkins, A. S. (1999), "The Web as a graph: measurements, models and methods," *Lecture Notes in Computer Science* **1627**, 1–18.

Klemm, K. and Eguíluz, V. M. (2002a), "Growing scale-free networks with small-world behavior," *Phys. Rev. E* **65**, 057102.

— (2002b), "Highly clustered scale-free networks," *Phys. Rev. E* **65**, 036123.

Krapivsky, P. L. and Redner, S. (2002), "A statistical physics perspective of Web growth," *Computer Networks* **39**, 261–276.

Krapivsky, P. L., Redner, S., and Leyvraz, F. (2000), "Connectivity of growing random networks," *Phys. Rev. Lett.* **85**, 4629.

Krapivsky, P. L., Rodgers, G., and Redner, S. (2001), "Degree distributions of growing networks," *Phys. Rev. Lett.* **86**, 5401.

Krzywicki, A. (2001), "Defining statistical ensembles of random graphs," e-print cond-mat/0110574.

Kumar, R., Raghavan, P., Rajagopalan, S., Sivakumar, D., Tomkins, A., and Upfal, E. (2000), "Stochastic models for the Web graph," in Proceedings of th 41th IEEE Symposium on Foundations of Computer Science (FOCS), pp. 57–65.

Kumar, R., Raghavan, P., Rajagopalan, S., and Tomkins, A. (1999), "Trawling the Web for emerging cyber-communities," *Computer Networks* (Amsterdam, Netherlands) **31**(11–16), 1481–1493.

Labovitz, C., Ahuja, A., Bose, A., and Jahanian, F. (2001), "Delayed Internet routing convergence," *IEEE-ACM Transactions on Networking* **9**, 293–306.

Labovitz, C., Ahuja, A., and Jahanian, F. (1999), "Experimental study of internet stability and wide-area network failures," in Proceedings of the IEEE Symposium on Fault-Tolerant Computing (FTCS99), Wisconsin.

Labovitz, C., Ahuja, A., Wattenhofer, R., and Srinivasan, V. (2001), "The impact of internet policy and topology on delayed routing convergence," in INFOCOM, pp. 537–546.

Labovitz, C., Malan, G. R., and Jahanian, F. (1998), "Internet routing instability," *IEEE-ACM Transactions on Networking* **6**, 515–528.

Lakhina, A., Byers, J. W., Crovella, M., and Matta, I. (2002), "On the geographic location of Internet resources," Proceedings of the ACM SIGCOMM Internet Measurement Workshop, Marseilles.

Lakhina, A., Byers, J. W., Crovella, M., and Xie, P. (2002), "Sampling biases in IP topology measurements," Technical report BUCS-TR-2002-021, Department of Computer Sciences, Boston University.

Latora, V. and Marchiori, M. (2001), "Efficient behavior of small-world networks," *Phys. Rev. Lett.* **87**, 198701.

Laura, L., Leonardi, S., Caldarelli, G., and De Los Rios, P. (2002), "A multi-layer model for the Web graph," in 2nd International Workshop on Web Dynamics.

Laura, L., Leonardi, S., Millozzi, S., Meyer, U., and Sibeyn, J. F. (2003), "Algorithms and experiments for the Webgraph," European Symposium on Algorithms.

Lawrence, S. and Giles, C. L. (1998), "Searching the World-Wide-Web," *Science* **280**, 98–100.

— (1999), "Accessibility of information on the Web," *Nature* **400**, 107–109.

Lee, C. and Stepanek, J. (2001), "On future global grid communication performance," 10th IEEE Heterogeneous Computing Workshop, May 2001.

Leland, W. E., Taqqu, M. S., Willinger, W., and Wilson, D. V. (1994), "On the self-similar nature of Ethernet traffic," *IEEE/ACM Transaction on Networking* **2**, 1–15.

Magnasco, M. O. (2000), "The thunder of distant Net storms," e-print nlin.AO/0010051.

Magoni, D. and Pansiot, J.-J. (2001), "Analysis of the autonomous system network topology," *ACM SIGCOMM Computer Communication Review* **31**(3), 26–37.

Mandelbrot, B. B. (1969), "Long run linearity, locally gaussian processes, H-spectra and infinite variances," *Intern. Econom. Rev.* **10**, 82–113.

— (1982), *The Fractal Geometry of Nature*, San Francisco: Freeman.

Mandelbrot, M. (1997), *Fractals and Scaling in Finance: Discontinuity and Concentration*, New York: Springer-Verlag.

Mantegna, R. and Stanley, H. E. (1999), *An Introduction to Econophysics: Correlations and Complexity in Finance*, Cambridge: Cambridge University Press.

Marendy, P. (2001), "A review of World Wide Web searching techniques, focusing on HITS and related algorithms that utilise the link topology of the World-Wide-Web to

provide the basis for a structure based search technology,"
citeseer.nj.nec.com/559198.html

Marro, J. and Dickman, R. (1999), *Nonequilibrium Phase Transitions in Lattice Models*, Cambridge: Cambridge University Press.

May, R. M. and Lloyd, A. L. (2001), "Infection dynamics on scale-free networks," *Phys. Rev. E* **64**, 066112.

McDonald, D. K. C. (1962), *Noise and Fluctuations*, New York: Wiley.

McGregor, A., Braun, H.-W., and Brown, J. (2000), "The {NLANR} network analysis infrastructure," *IEEE Communications Magazine* **38**, 122–129.

Medina, A., Matta, I., and Byers, J. (2000), "On the origin of power laws in Internet topologies," *Comput. Commun. Rev.* **30**, 18–28.

Menczer, F. (2002), "Growing and navigating the small world Web by local content," *Proc. Natl. Acad. Sci. USA* **99**, 14014–14019.

— (2003), "Complementing search engines with online Web mining agents," *Decision Support Systems* **35**, 195–212.

Menczer, F. and Belew, R. (2000), "Adaptive retrieval agents: Internalizing local context and scaling up to the Web," *Machine Learning Journal* **39**, 203–242.

Milgram, S. (1967), "The small world problem," *Psychology Today* **2**, 60–67.

Molloy, M. and Reed, B. (1995), "A critical point for random graphs with a given degree sequence," *Random Struct. Algorithms* **6**, 161.

— (1998), "The size of the giant component of a random graph with a given degree sequence," *Combinatorics, Probab. Comput.* **7**, 295.

Moore, C. and Newman, M. E. J. (2000), "Epidemics and percolation in small-world networks," *Phys. Rev. E* **61**, 5678.

Moore, D., Shannon, C., and Brown, J. (2002), "Code-Red: a case study on the spread and victims of an Internet worm," Proceedings of the 2nd Internet Measurement Workshop.

Moreno, Y., Pastor-Satorras, R., and Vespignani, A. (2002), "Epidemic outbreaks in complex heterogeneous networks," *Eur. Phys. J. B* **26**, 521–529.

Moreno, Y., Pastor-Satorras, R., Vazquez, A., and Vespignani, A. (2003), "Critical load and congestion instabilities in scale-free networks," *Europhys. Lett.* **62**, 292–298.

Moreno, Y. and Vázquez, A. (2003), "Disease spreading in structured scale-free networks," *Eur. Phys. J. B* **31**, 265–271.

Morton, D. (1997), "Understanding IPv6," *PC Network Advisor* **83**, 17–22.

Mossa, S., Barthélémy, M., Stanley, H. E., and Amaral, L. A. N. (2002), "Truncation of power law behavior in 'scale-free' network models due to information filtering," *Phys. Rev. Lett.* **88**, 138701.

Motter, A. E. and Lai, Y.-C. (2002), "Cascade-based attacks on complex networks," *Phys. Rev. E* **66**, 065102.

Murray, J. D. (1993), *Mathematical Biology*, 2nd edn, Berlin: Springer Verlag.

Murray, M. and Claffy, K. (2001), "Measuring the immeasurable: global Internet measurement infrastructure," in PAM2001 – A workshop on passive and active measurements, RIPE NCC, Amsterdam, Netherlands.

Murray, W. H. (1988), "The application of epidemiology to computer viruses," *Computers and Security* **7**, 130–150.

Newman, M. E. J. (2001a), "Who is the best connected scientist? A study of scientific coauthorship networks I," *Phys. Rev. E* **64**, 016131.

— (2001b), "Who is the best connected scientist? A study of scientific coauthorship networks II," *Phys. Rev. E* **64**, 016132.

— (2002a), "Assortative mixing in networks," *Phys. Rev. Lett.* **89**, 208701.

— (2002b), "Spread of epidemic diseases on networks," *Phys. Rev. E* **64**, 016128.

— (2003), "Random graphs as models of networks," in S. Bornholdt and H. G. Schuster
   (eds.), *Handbook of Graphs and Networks: From the Genome to the Internet*, Berlin:
   Wiley-VCH, pp. 35–68.

Newman, M. E. J., Forrest, S., and Balthrop, J. (2002), "E-mail networks and the spread
   of computer viruses," *Phys. Rev. E* **66**, 035101.

Newman, M. E. J., Strogatz, S. H., and Watts, D. J. (2001), "Random graphs with
   arbitrary degree distribution and their applications," *Phys. Rev. E* **64**, 026118.

Ohira, T. and Sawatari, R. (1998), "Phase transition in a computer network traffic model,"
   Phys. Rev. E **58**, 193–195.

Pansiot, J.-J. and Grad, D. (1998), "On routes and multicast trees in the Internet," *ACM
   Comp. Comm. Rev.* **28**, 41.

Pareto, V. (1896), *Course d'Economie Politique*, Geneve: Droz.

Park, K., Kim, G., and Crovella, M. (1996), "On the relationship between file sizes,
   transport protocols, and self-similar network traffic," in Proc. IEEE International
   Conference on Network Protocols, pp. 171–180.

— (1997), "On the effect of traffic self-similarity on network performance," in Proc. SPIE
   International Conference on Performance and Control of Network Systems,
   pp. 296–310.

Park, K. and Lee, H. (2001), "On the effectiveness of route-based packet filtering for
   distributed DoS attack prevention in power-law internets," in Proc. of ACM
   SIGCOMM, pp. 15–26.

Park, S.-T., Khrabrov, A., Pennock, D. M., Lawrence, S., Giles, C. L., and Ungar, L. H.
   (2003), "Static and dynamic analysis of the internet's susceptibility to faults and
   attacks," in The 22nd Annual Joint Conference of the IEEE Computer and
   Communications Societies (INFOCOM 2003), San Francisco.

Pastor-Satorras, R., Vázquez, A., and Vespignani, A. (2001), "Dynamical and correlation
   properties of the Internet," *Phys. Rev. Lett.* **87**, 258701.

Pastor-Satorras, R. and Vespignani, A. (2001a), "Epidemic dynamics and endemic states
   in complex networks," *Phys. Rev. E* **63**, 066117.

— (2001b), "Epidemic spreading in scale-free networks," *Phys. Rev. Lett.* **86**, 3200–3203.

— (2001c), "Immunization of complex networks," *Phys. Rev. E* **65**, 036104.

Pastor-Satorras, R. and Vespignani, A. (2002), "Epidemic dynamics in finite size scale-
   free networks," *Phys. Rev. E* **65**, 035108.

Pathria, R. K. (1996), *Statistical Mechanics*, 2nd edn, Oxford: Butterworth-Heinemann.

Paxson, V. (1997), "End-to-end routing behavior in the Internet," *IEEE/ACM
   Transactions on Networking* **5**, 601–615.

Paxson, V. and Floyd, S. (1995), "Wide area traffic:the failure of poisson modeling,"
   IEEE/ACM Transaction on Networking **3**, 226–244.

Pennock, D. M., Flake, G. W., Lawrence, S., Glover, E. J., and Giles, C. L. (2002),
   "Winners don't take all: characterizing the competition for links on the Web," *Proc.
   Natl. Acad. Sci. USA* **99**, 5207–5211.

Percacci, R. and Vespignani, A. (2003), "Scale-free behavior of the Internet global
   performance," *Eur. Phys. J. B* **32**, 411–414.

Qian, C., Chang, H., Govindan, R., Jamin, S., Shenker, S., and Willinger, W. (2002), "The
   origin of power laws in Internet topologies revisited," in *INFOCOM 2002,
   Twenty-First Annual Joint Conference of the IEEE Computer and Communications
   Societies, Proceedings IEEE*, Volume 2, IEEE Computer Society Press, pp. 608–
   617.

Ravasz, E. and Barabási, A.-L. (2003), "Hierarchical organization in complex networks,"
   Phys. Rev. E **67**, 026112.

Redner, S. (1998), "How popular is your paper? An empirical study of citation
   distribution," *Eur. Phys. J. B* **4**, 131.

Ripeanu, M., Foster, I., and Iamnitchi, A. (2002), "Mapping the gnutella network: Properties of large-scale peer-to-peer systems and implications for system design," *IEEE Internet Computing Journal* **6**, 50–57.

Rodriguez-Iturbe, I. and Rinaldo, A. (1997), *Fractal River Basins: Chance and Self-Organization*, Cambridge: Cambridge University Press.

Russel, S. and Norvig, P. (1995), *Artificial Intelligence: A Modern Approach*, New York: Prentice Hall.

Saroiu, S., Gummadi, P. K., and Gribble, S. D. (2002), "A measurement study of peer-to-peer file sharing systems," in Proceedings of Multimedia Computing and Networking 2002 (MMCN 2002), San Jose, CA, USA.

Sarshar, N. and Roychowdhury, V. (2003), "Scale-free and stable structures in *ad hoc* networks," e-print cond-mat/0303041.

Schwartz, I. B. and Billings, L. (2003), "Dimacs Working Group Report on analogies between computer and biological viruses and their immune systems," http://dimacs. rutgers.edu/Workshops/Analogies/index.html

Smith, R. D. (2002), "Instant messaging as a scale-free network," e-print cond-mat/0206378.

Solé, R. V. and Valverde, S. (2001), "Information transfer and phase transitions in a model of Internet traffic," *Physica A* **289**, 595–605.

Staniford, S., Paxson, V., and Weaver, N. (2002), "How to own the Internet in your spare time," in the Proceedings of the 11th USENIX Security Symposium (Security '02).

Stanley, H. E. (1971), *Introduction to Phase Transitions and Critical Phenomena*, Oxford: Oxford University Press.

Stauffer, D. and Aharony, A. (1994), *Introduction to Percolation Theory*, 2nd edn, London: Taylor & Francis.

Strogatz, S. H. (2001), "Exploring complex networks," *Nature* **410**, 268–276.

Suplee, C. (2000), "Anatomy of a 'love bug,'" *Washigton Post*.

Tadic, B. (2001), "Dynamics of directed graphs: the World Wide Web," *Physica A* **293**, 273.

Takayasu, M., Fukuda, K., and Takayasu, H. (1999), "Application of statistical physics to the Internet traffics," *Physica A* **274**, 140–148.

Tangmunarunkit, H., Doyle, J., Govindan, R., Jamin, S., Shenker, S., and Willinger, W. (2001), "Does AS size determine degree in AS topology?" *Comput. Commun. Rev.* **31**, 7–10.

Tangmunarunkit, H., Govindan, R., Jamin, S., Shenker, S., and Willinger, W. (2002a), "Network topologies, power laws, and hierarchy," *Comput. Commun. Rev.* **32**, 76–76.

— (2002b), "Network topology generators: Degree-based vs. structural," http://citeseer. nj.nec.com/531065.html

Uhlig, S. and Bonaventure, O. (2001), "The macroscopic behavior of Internet traffic: a comparative study," Technical Report Infonet-TR-2001-10, University of Namur.

Valverde, S. and Solé, R. V. (2002), "Self-organized critical traffic in parallel computer networks," *Physica A* **312**, 636–648.

Vázquez, A. (2001), "Knowing a network by walking on it: emergence of scaling," *Europhys. Lett.* **54**, 430.

— (2002), "Growing networks with local rules: preferential attachment, clustering hierarchy and degree correlations," e-print cond-mat/0211528.

Vázquez, A. and Moreno, Y. (2003), "Resilience to damage of graphs with degree correlations," *Phys. Rev. E* **67**, 015101.

Vázquez, A., Pastor-Satorras, R., and Vespignani, A. (2002a), "Internet topology at the router and autonomous system level," e-print cond-mat/0206084.

— (2002b), "Large-scale topological and dynamical properties of Internet," *Phys. Rev. E* **65**, 066130.

— (2003), "Reconstructing networks topology by random path probing," LPT preprint.

Veres, A., Kenesi, Z., Molnár, S., and Vattay, G. (2000), "On the Propagation of Long-Range Dependence in the Internet," in ACM SIGCOMM, Stockholm, Sweden, pp. 26–33.

Vukadinovic, D., Huang, P., and Erlebach, T. (2002), "On the spectrum and structure of Internet topology graphs," in H. U. (ed.) *Proceedings of the Innovative Internet Computing Systems Workshop: Lecture Notes in Computer Science*, Springer-Verlag, pp. 83–95.

Wasserman, S. and Faust, K. (1994), *Social Network Analysis: Methods and Applications*, Cambridge: Cambridge University Press.

Watts, D. J. (1999), *Small Worlds: The Dynamics of Networks between Order and Randomness*, New Jersey: Princeton University Press.

— (2003), *Six Degrees: The Science of a Connected Age*, New York: W. W. Norton & Company.

Watts, D. J., Dodds, P. S., and Newman, M. E. J. (2002), "Identity and search in social networks," *Science* **296**, 1302–1305.

Watts, D. J. and Strogatz, S. H. (1998), "Collective dynamics of "small-world" networks," *Nature* **393**, 440–442.

Waxman, B. M. (1988), "Routing of multipoint connections," *IEEE Journal of Selected Areas in Communications* **6**, 1617–1622.

Wei, L. and Estrin, D. (1994), "The trade-offs of multicast trees and algorithms," in Proceedings of ICCCN.

White, S. R. (1998), "Open problems in computer virus research," Virus Bulletin Conference, Munich.

Willinger, W., Govindan, R., Jamin, S., Paxson, V., and Shenker, S. (2002), "Scaling phenomena in the Internet: critically examining criticality," *Proc. Natl. Acad. Sci. USA* **99**, 2573–2580.

Willinger, W., Taqqu, M. S., and Erramilli, A. (1996), "A bibliographic guide to self-similar traffic and performance modeling for modern high-speed networks," in F. Kelly, S. Zachary, and I. Ziedins (eds.), "*Stochastic Networks: Theory and Applications*," Oxford: Clarendon Press. pp. 82–89.

Yang, B. and Garcia-Molina, H. (2002), "Improving search in peer-to-peer networks," in 22nd International Conference on Distributed Computing Systems (ICDCS 2002), Vienna, pp. 5–14.

Yeomans, J. M. (1992), *Statistical Mechanics of Phase Transitions*, Oxford: Oxford University Press.

Yook, S.-H., Jeong, H., and Barabási, A.-L. (2002), "Modeling the Internet's large-scale topology," *Proc. Nat. Acad. Sci. USA* **99**, 13382–13386.

Zegura, E. W., Calvert, K., and Donahoo, M. J. (1997), "A quantitative comparison of graph-based models for Internet topology," *IEEE/ACM T. Network.* **5**, 770.

Zegura, E. W., Calvert, K. L., and Bhattacharjee, S. (1996), "How to model an Internetwork," in IEEE Infocom, Vol. 2, IEEE, San Francisco, CA, pp. 594–602.

Zhou, S. and Mondragon, R. J. (2003), "The missing links in the BGP-based AS connectivity maps," in Proceedings of Passive and Active Measurements (PAM2003).

Zipf, G. (1949), *Human Behavior and the Principle of Least Effort*, Cambridge, MA: Addison-Wesley.

Index

Printed in the United States
By Bookmasters